SpringerBriefs in Applied Sciences and Technology

For further volumes:
http://www.springer.com/series/8884

Pratima Bajpai

Advances in Bioethanol

 Springer

Pratima Bajpai
Thapar Centre for Industrial Research
    and Development
Patiala
India

ISSN 2191-530X   ISSN 2191-5318   (electronic)
ISBN 978-81-322-1583-7   ISBN 978-81-322-1584-4   (eBook)
DOI 10.1007/978-81-322-1584-4
Springer New Delhi Heidelberg New York Dordrecht London

Library of Congress Control Number: 2013945842

Printed on acid-free paper

Springer is part of Springer Science+Business Media (www.springer.com)

Preface

Disadvantages of fossil fuel derived transportation fuels (greenhouse gas emissions, pollution, resource depletion, unbalanced supply-demand relations) are strongly reduced or even absent with biotransportation fuels. Of all biofuels, ethanol is already produced on a fair scale. It produces slightly less greenhouse emissions than fossil fuel (carbon dioxide is recycled from the atmosphere to produce biomass); can replace harmful fuel additives (e.g., methyl tertiary butyl ether) and produces jobs for farmers and refinery workers. It is easily applicable in present day internal combustion engine vehicles (ICEVs), as mixing with gasoline is possible. Ethanol is already commonly used in a 10 % ethanol/90 % gasoline blend. Adapted ICEVs can use a blend of 85 % ethanol/15 % gasoline (E85) or even 95 % ethanol (E95). Ethanol addition increases octane and reduces carbonmonoxide, volatile organic carbon and particulate emissions of gasoline. And, via on board reforming to hydrogen, ethanol is also suitable for use in future fuel cell vehicles (FCVs). Those vehicles are supposed to have about double the current ICEV fuel efficiency.

Ethanol production and use has spread to every corner of the globe. As concerns over petroleum supplies and global warming continue to grow, more nations are looking to ethanol and renewable fuels as a way to counter oil dependency and environmental impacts. World production reached an all-time high of nearly 23 billion gallons in 2010 and is expected to exceed 1,20,000 million mark by the end of the year 2020. While the US became the world's largest producer of fuel ethanol in 2010, Brazil remains a close second, and China, India, Thailand and other nations are rapidly expanding their own domestic ethanol industries. Increased production and use of ethanol have also led to a growing international trade for the renewable fuel. While the vast majority of ethanol is consumed in the country in which it is produced, some nations are finding it more profitable to export ethanol to countries like the US and Japan. High spot market prices for ethanol and the rapid elimination of MTBE by gasoline refiners led to record imports into the US in the last few years. More than 500 million gallons of ethanol entered through American ports, paid the necessary duties, and competed effectively in the marketplace. The increased trade of ethanol around the world is helping to open up new markets for all sources of ethanol.

The sustainable production of bioethanol requires well planned and reasoned development programs to assure that the many environmental, social and economic concerns related to its use are addressed adequately. The key for making ethanol competitive as an alternative fuel is the ability to produce it from low-cost biomass. Many countries around the world are working extensively to develop new technologies for ethanol production from biomass, from which the lignocellulosic materials conversion seem to be the most promising one. This e-book provides an updated and detailed overview on Advances in Bioethanol. It looks at the historical perspectives, chemistry, sources and production of ethanol and discusses biotechnology breakthroughs and promising developments, its uses, advantages, problems, environmental effects and characteristics. In addition, it presents information about ethanol in different parts of the world and also highlights the challenges and future of ethanol.

Contents

Chapter 1
Introduction

1.1 Background

Energy consumption has increased steadily over the last century as the world population has grown and more countries have become industrialized. Crude oil has been the major resource to meet the increased energy demand. Campbell and Laherrere (1998) used several techniques to estimate the current known crude oil reserves and the reserves as yet undiscovered. They predicted that annual global oil production would decline to approximately 5 billion barrels in 2050. Because the economy in the USA and many other nations depends on oil, the consequences of inadequate oil availability could be severe. So, there is a great interest in exploring alternative energy sources (Wyman 1990; Lynd 2004; Herrera 2004; Tanaka 2006; Dien et al. 2006; Sun and Cheng 2004; Yacobucci 2003; Chandel 2007; Gray et al. 2006; Kheshgi et al. 2000; DOE 2007a; Badger 2002; EBIO 2006; Bajpai 2007). The use of ethanol as an alternative motor fuel has been steadily increasing around the world (BP 2006; Jessel 2006; Moreira and Walter 2005). Domestic production and use of ethanol for fuel can decrease dependence on foreign oil, reduce trade deficits, create jobs in rural areas, reduce air pollution, and reduce global climate change carbon dioxide buildup. Ethanol, unlike gasoline, is an oxygenated fuel that contains 35 % oxygen, which reduces particulate and NOx emissions from combustion. Unlike fossil fuels, ethanol is a renewable energy source produced through fermentation of sugars.

The term biofuel is attributed to any alternative fuel that derives from organic material, such as energy crops (corn, wheat, sugarcane, sugar beet, cassava, among others), crop residues (e.g., rice straw, rice husk, corn stover, corn cobs), or waste biomass (for instance, food waste, livestock waste, paper waste, construction-derived wood residues, and others). Of all biofuels, ethanol has been trusted as an alternate fuel for the future and is already produced on a fair scale worldwide.

Some excerpts taken from Bajpai (2007). PIRA Technology Report on Bioethanol with kind permission from Smithers PIRA.

The bulk of the production is located in Brazil (ANP 2007) and the USA (Hamelinck et al. 2005). In this sense, bioethanol is expected to be one of the dominating renewable biofuels in the transport sector within the coming 20 years (Hägerdal et al. 2006; Berg 2004; Paszner 2006). In 2010, almost 90 % of bioethanol being used for fuel is being used in USA and Brazil. It can be utilized as a liquid fuel in internal combustion engines, either neat or in blends with petroleum (Table 1.1). Table 1.2 compares the energy content of bioethanol with conventional fossil fuels used for road and aviation transportation. In Brazil, ethanol blends are mandatory (E20–E25) and anhydrous ethanol (E100) is also available from thousands of filling stations. In addition, there are 6 million flex-fuel vehicles in Brazil and 3 million able to run on E100. Bioethanol now accounts for ~50 % of the Brazilian transport fuel market, where gasoline may now be regarded as the "alternative" fuel. In Brazil, >20 % of cars (and some light aircraft) are able to use E100 (100 % ethanol) as fuel, which includes ethanol-only engines and flex-fuel vehicles which are able to run with either neat ethanol, neat gasoline, or any mixture of both (Lynch 2006).

Today, the main reason for the interest in renewable biofuels is the possibility of obtaining a substantial reduction in noxious exhaust emissions from combustion, especially as statutory limits are becoming more stringent and more exhaust components are regulated. Wider use of a chemically simple fuel such as bioethanol will mean that there are fewer harmful effects on life and ecosystems. Using ethanol in place of gasoline helps to reduce carbon dioxide (CO_2) emissions by up

Table 1.1 Bioethanol-gasoline blends* used in different countries

Country	Blend
Europe	E5 Common in unleaded petrols
	E85 Relatively uncommon at present
USA	E10 10 % Ethanol in gasoline is common
	(gasohol)
	E70–E85 Blend varies with state
Brazil	E25–E75 Higher blends possible via flex-fuel vehicles
	E100

*(E = ethanol and number represents % in gasoline)
Based on Walker (2010)

Table 1.2 Energy content of fossil fuels and bioethanol

Fuel	Energy content, MJ/L
Gasoline (regular)	34.8
Gasoline (aviation)	33.5
Diesel	38.6
Autogas (LPG)	26.8
E100	23.5
E85	25.2
E10	33.7

Based on Walker (2010)

Table 1.3 Ethanol facts (environment)

Using ethanol in place of gasoline helps to reduce carbon dioxide (CO_2) emissions by up 30–50 % given today's technology.
In 2012, the 13.2 billion gallons of ethanol produced reduced greenhouse gas emissions from on-road vehicles by 33.4 million tons. That is equivalent to removing 5.2 million cars and pickups (comparable with the number of registered vehicles in the state of Michigan) from the road for one year.
New technologies are increasing ethanol yields, improving efficiencies and allowing ethanol biorefineries to make better use of natural resources.
Ethanol has a positive energy balance.
The American Lung Association supports the use of E85, a blend of 85 % ethanol and 15 % gasoline for flex-fuel vehicles specifically designed to operate on this fuel.
Ethanol reduces tailpipe carbon monoxide emissions by as much as 30 %, toxics content by 13 % (mass) and 21 % (potency), and tailpipe fine particulate matter (PM) emissions by 50 %.
Ethanol is the oxygenate of choice in the federal winter oxygenated fuels program in cities that exceed public health standards for carbon monoxide pollution.
Ethanol is rapidly biodegraded in surface water, groundwater and soil, and is the safest component in gasoline today.
Ethanol reduces smog pollution.
Water usage in ethanol production is declining.

30–50 % given today's technology (Table 1.3). In particular, people living in urban areas may in future appreciate the use of improved low-emission vehicles that have an agreeable smell, are smokeless, and are propelled either by reformulated bioethanol, by bioethanol blended with gasoline, or by neat biofuels. How the air quality can be improved is something that is increasingly worth investigating, for the sake of people and the environment! Large-scale, sustainable, worldwide production and use of bioethanol from biomass resources will produce tangible significant benefits for our growing and fast-evolving society, as well as for the earth's climate. Important environmental benefits could be achieved in the socio-economic development of large rural populations and the diversification of energy supply, in particular for the strategically vital sector of transport (Turkenburg 2000). A life-cycle analysis of ethanol production—from field to the car—by the US Department of Agriculture found that ethanol has a large and positive energy balance. Ethanol yields 134 % of the energy used to grow and harvest the corn and process into ethanol. By comparison, gasoline yields only 80 % of the energy used to produce it. Bioethanol does not add to global CO_2 levels because it only "recycles" CO_2 already present in the atmosphere. More specifically, CO_2 is removed from the atmosphere through photosynthesis when crops intended for conversion to bioethanol are grown (Fig. 1.1). CO_2 is then released (returned) to the atmosphere during combustion. In contrast, burning a fossil fuel such as petrol adds to global CO_2 because it releases new amounts of CO_2 that were previously trapped underground for millions of years. Finally, unlike oil, bioethanol is a renewable fuel, which inherently helps the environment by allowing us to conserve other energy resources (EIA 1999, 2006).

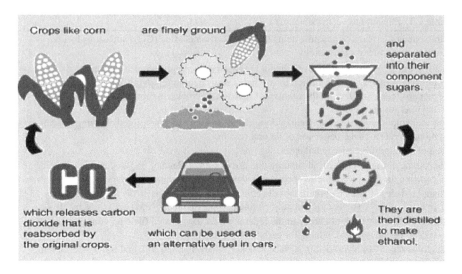

Fig. 1.1 Fuel ethanol process. Based on Launder (1999)

The advantages and disadvantages of ethanol fuel are detailed in Table 1.4. Ethanol is a flexible transportation fuel that can be used in anhydrous form at 99.6 Gay Lussac (GL) as a blending agent in ethanol–gasoline blends, either directly or indirectly or as a primary fuel in neat hydrous form (95.5 GL). Ethanol makes an excellent motor fuel: It has a research octane number (RON) of 109 and a motor octane number (MON) of 90, which are higher compared to gasoline, which has a RON of 91–98 and a MON of 83–90. Ethanol also has a lower vapor pressure than gasoline (its Reid vapor pressure is 16 kPa versus 71 for typical gasoline), resulting in less evaporative emissions. Ethanol's flammability in air (1.3–7.6 % v/v) is also much lower compared to gasoline (3.5–19 % v/v), reducing the number and severity of vehicle fires (Goldemberg et al. 1993). On the other hand, when used as a neat fuel, ethanol has a lower energy density than gasoline (ethanol has lower and higher heating values of 21.2 and 23.4 MJ/L, respectively; for gasoline, the values are 30.1 and 34.9 MJ/L) and cold-start problems exist as well (McMillan 1997).

Ethanol is also suitable for use in future fuel cell vehicles (FCVs) (DOE 2007b). Those vehicles are supposed to have about double the current ICEV fuel efficiency (Lynd 1996). Beginning with the model year 1999, an increasing number of vehicles in the world are manufactured with engines which can run on any gasoline from 0 % ethanol up to 85 % ethanol without modification. Many light trucks are designed to be dual-fuel or flexible-fuel vehicles, since they can automatically detect the type of fuel and change the engine's behavior, principally air-to-fuel ratio and ignition timing to compensate for the different octane levels of the fuel in the engine cylinders.

Ethanol has three major uses: as a renewable fuel, as a beverage, and for industrial purposes (Table 1.5). Of the three grades of ethanol, fuel-grade ethanol is

Table 1.4 Advantages and disadvantages of ethanol fuel

Advantages

Renewable, relatively safe fuel that can be used with few engine modifications.

Can improve agricultural economies by providing farmers with a stable market for certain crops, such as maize and sugar beets.

Increases national energy security because some use of foreign petroleum is averted.

Non-toxic and biodegradable, quickly breaks down into harmless substances if spilled. Ethanol use reduces carbon monoxide and many toxic pollutants from the tailpipe of vehicles, making air cleaner.

As it is made from crops that absorb carbon dioxide and give off oxygen, it has a potential to reduce greenhouse gas emission and help maintain the balance of carbon dioxide in the atmosphere.

Ethanol, unlike petroleum, is claimed to be a form of renewable energy that can be produced from agricultural crops such as sugarcane, potato, and corn It has many other advantages over gasoline at least it can be a permanent resource.

Disadvantages

Ethanol has a lower heat of combustion (per mole, per unit of volume, and per unit of mass) that petroleum

Large amounts of arable land are required to produce the crops required to obtain ethanol, leading to problems such as soil erosion, deforestation, fertilizer runoff and salinity

Major environmental problems would arise out of the disposal of waste fermentation liquors

Ethanol powered vehicles will have trouble starting up at low temperatures.

Typical current engines would require modification to use high concentrations of ethanol

Harder to transport

Based on http://www.wisegeek.org/what-are-the-advantages-and-disadvantages-of-ethanol-fuel.htm
http://www.solarpowernotes.com/renewable-energy/ethanol-fuel.html

driving record ethanol production in many countries. About 95 % of all ethanol is derived by fermentation from sugar or starch crops; the rest is produced synthetically. The synthesis route evolves dehydration of hydrocarbons (e.g., ethylene) or by reaction with sulfuric acid, to produce ethyl sulfate, followed by hydrolysis. The production routes from biomass are based on fermentation or hydrolysis. According to F.O. Licht, synthetic alcohol production is concentrated in the hands of a few mostly multinational companies such as Sasol, with operations in South Africa and Germany, SADAF of Saudi Arabia, a 50:50 joint venture between Shell of the United Kingdom (UK) and the Netherlands, the Saudi Arabian Basic Industries Corporation, BP of the UK, as well as Equistar in the United States.

Fermentation ethanol is mainly produced for fuel, though a small share is used by the beverage industry and the industrial industry. The bulk of the production and consumption is located in Brazil and the USA. Fermentation technologies for sugar and starch crops are very well developed, but have certain limits—these crops have a high value for food application, and their sugar yield per hectare is very low compared with the most prevalent forms of sugar in nature: cellulose and hemicellulose. Suitable processes for lignocellulosic biomass therefore have room for much further development: A bigger crop variety can be employed, a larger portion of these crops can be converted, and hence, larger scales and lower

Table 1.5 Major uses of ethanol

Beverage alcohol
Ethanol is often called "drinking alcohol" as it is the prime ingredient in alcoholic beverages. Ethanol is the intoxicating substance in alcohol.

Fuel alcohol
Ethanol can be used as a fuel for motor vehicles. Ethanol makes a good fuel for cars because it reduces the emission of harmful gases such as carbon monoxide. Brazil are one of the leaders in the production of cars that run on ethanol. Over 20 % of cars in Brazil are able to run on 100 % ethanol fuel.
Lightweight rocket-powered racing aircraft often use ethanol as rocket fuel.
Ethanol is used in antiseptic and some antibacterial soaps and wipes. Ethanol is effective against viruses, fungi, and most bacteria but is ineffective against bacterial spores.

Industrial alcohol
As ethanol is soluble in water, it can be used in a variety of different products. These include paint, permanent markers, perfumes, and deodorants. Ethanol may also be used as a solvent in cooking, such as vodka sauce.
Ethanol is considered a "feedstock" into the chemical industry as it is used to make other important chemicals.

Based on Berg (2004)

costs are possible. There is copious amount of lignocellulosic biomass worldwide that can be exploited for fuel ethanol production. According to US Department of Energy, cellulosic ethanol reduces greenhouse gas emissions by 85 % over reformulated gasoline. By contrast, sugar-fermented ethanol reduces greenhouse gas emissions by 18–19 % over gasoline. Beyond added environmental benefits, cellulose-based ethanol could offer additional revenue streams to farmers for the collection and sale of currently unused corn stover or straw, for example. Global production of biofuels increased 17 % in 2010 to reach an all-time high of 105 billion liters (28 billion gallons US), up from 90 billion liters (24 billion gallons US) in 2009. High oil prices, a global economic rebound, and new laws and mandates in Argentina, Brazil, Canada, China, and the United States, among other countries, are all factors behind the surge in production. Biofuels provided 2.7 % of all global fuel for road transportation—an increase from 2 % in 2009. The two biofuel alternatives to fossil fuels for transportation largely consist of ethanol and biodiesel. The world produced 86 billion liters (23 billion gallons US) of ethanol in 2010, 18 % more than in 2009. World biodiesel production rose to 19 billion liters (5 billion gallons US) in 2010, a 12 % increase from 2009. The United States and Brazil remain the two largest producers of ethanol. In 2010, the United States generated 49 billion liters (13 billion gallons US), or 57 % of global output, and Brazil produced 28 billion liters (7 billion gallons US), or 33 % of the total. Corn is the primary feedstock for US ethanol, and sugarcane is the dominant source of ethanol in Brazil. In the United States, the record production of biofuels is attributed in part to high oil prices, which encouraged several large fuel companies, including Sunoco, Valero, Flint Hills, and Murphy Oil, to enter the ethanol industry. High oil prices were also a factor in Brazil, where every third car owner drives a flex-fuel vehicle that can run on either fossil or bio-based fuels. Although the USA and Brazil are the world leaders in ethanol, the largest producer of biodiesel is the European Union, which generated 53 % of all biodiesel in 2010. However, some European countries may switch from biodiesel to ethanol because a recent report from the European Commission states that ethanol crops have a higher energy content than biodiesel crops, making them more efficient sources of fuel.

1.2 Historical Perspectives

In 1925, Henry Ford had quoted ethyl alcohol, ethanol, as "the fuel of the future." He furthermore stated, "The fuel of the future is going to come from apples, weeds, sawdust—almost anything. There is fuel in every bit of vegetable matter that can be fermented." Today, Henry Ford's futuristic vision significance can be easily understood.

In ancient times, ethanol was known as an intoxicating drink. It is the same alcohol used in beverage alcohol but meets fuel-grade standards. Ethanol that is to be used as a fuel is "denatured" by adding a small amount of gasoline to it. This makes it unfit for drinking. During the late 1800s, ethanol was used in the United States for lamp fuel and sales exceeded 25 million gallons per year (Morris 1993). At the request of large oil companies, the government placed a tax on ethanol during the Civil War. This tax almost destroyed the ethanol industry. In 1906, the tax was lifted and alcohol fuel did well until competition from oil companies greatly reduced its use. The first large-scale use of ethanol as a fuel occurred during the early 1900s when petroleum supplies in Europe were short. In America, Henry Ford's Model T and other early 1920s automobiles were originally designed to run on alcohol fuels. Germany and the USA both relied on ethanol to power vehicles for their armies during World War II. After World War II, oil prices decreased which caused the use of ethanol to decrease as well. The limited use of ethanol continued until the oil crisis in the early 1970s. The use of ethanol as a fuel has grown since the late 1970s. It was first used as a gasoline extender because of oil shortages. In 1973, the Organization of Petroleum Exporting Countries (OPEC) caused gasoline shortages by increasing prices and blocking shipments of crude oil to the United States. The OPEC action called attention to the fact that the United States was extremely dependent on foreign oil. The focus shifted once again to alternative fuels such as ethanol. At that time, gasoline containing ethanol was called "gasohol." Later, when gasoline was more plentiful, ethanol-blended gasoline was introduced to increase the octane rating and the name "gasohol" was dropped in favor of names reflecting the higher octane levels. "E-10 Unleaded" and "super unleaded" are examples of names used today.

Ever since the modest inception of a sizable ethanol industry, technology on the whole has risen, thus developing lower cost methods of producing greater quantities of fuel ethanol which are simultaneously more efficient in their use of fossil fuel inputs. These combined effects have helped production of ethanol fuel rise in US by more than 225 % between 2001 and 2005 (Renewable Fuels Association 2006). Ethanol has also been used outside the United States, most notably in Brazil which started a program of government-mandated ethanol production in 1975 and has since encouraged production of flex-fuel vehicles (FFVs) and cars fueled entirely by ethanol (Luhnow and Samor 2006). Due in part to this jump start on ethanol production and its geographic advantage in growing sugarcane (an ideal ethanol feedstock), Brazil is one of the biggest producer of ethanol. Brazil is so efficient that it can produce a gallon of ethanol for about one dollar (Luhnow and Samor 2006). The Brazilian ethanol market, which was once dependent on governmental regulation and subsidies, has blossomed into a system that thrives even without regulation. Fuel ethanol production in the United States caught up to Brazil's for the first time, growing by 15 % in 2005, as both remained the dominant producers (REN21 2006). Although there are cultural and institutional differences between USA and Brazil, the general pattern of ethanol production and consumption under a regulatory environment in the USA could closely mirror what has happened in Brazil. Their policy effectiveness can be used as a benchmark for the American market.

1.3 Key Drivers and Trends

When evaluating key drivers for ethanol demand, energy security and climate change are considered to be the most important objectives reported by nearly all countries that engage in bioenergy development activities. The key drivers are as follows:

1. *Environmental*
2. *Energy security*
3. *Social and economic pressures*

Throughout the world, concern with clean air is a social and political priority; for example, the necessity of reducing pollutant emissions and achieving targets defined by Kyoto Protocol. Increasing dependency on imported energy supply, especially in a context of rising of oil prices, is also a general concern, particularly in the United States and European Union. Another most important driving force for ethanol production is generation of huge amount of new employment. Ethanol industry in Brazil is responsible for about one million direct jobs, approximately 50 % in the sugarcane production. Indirect jobs are estimated to be between 2.5 and 3.0 million. However, it should be mentioned that this high employment is partly due to low level of mechanization of agricultural activities, as well as poor automation at the industrial site (Rosillo-Calle and Walter 2006). World Bank reports that biofuel industries require about 100 times more workers per unit of energy produced than the fossil fuel industry. Transportation, including emissions from the production of transport fuels, is responsible for about one-quarter of energy-related greenhouse gas (GHG) emissions, and that share is rising. A US report (see http:/www.bio.org/EconomicImpactAdvancedBiofuels.pdf) has analyzed how growth of an advanced biofuel industry impacts on job creation, economic output, energy security, and investment opportunities. For example, biofuel industry could create 29,000 new jobs and $5.5 billion in economic growth over the next 3 years and could ultimately create 800,000 new jobs by 2022 with a positive effect on output of $148.7 billion. It is estimated that in the US, the cumulative total of avoided petroleum imports over the period 2010–2022 would exceed $350 billion. To stimulate the further development of US bioethanol, regulators should approve the deployment of E15 (15 % ethanol, 85 % gasoline) and to extend the tax credit for all ethanol feedstocks.

USDA released a report on February 11, 2013, detailing its agricultural projections for the next 10-year period. According to the paper, titled "USDA Agricultural Projections to 2022," long-run developments for global agriculture reflect steady world economic growth and continued global demand for biofuels. Combined, the report said these factors support longer-run increases in consumption, trade, and the price of agricultural products. Regarding biofuels, the report notes that demand for ethanol and biodiesel feedstocks is projected to continue growing, but at a slower pace than in recent years. The USDA also specified that expansion will continue to depend on biofuel policies. The dominance of the USA, Brazil, EU member countries, Argentina, Canada, China, and Indonesia in biofuel markets is expected to continue over the next 10 years. According to the USDA, these countries accounted for 90 % of world biofuel production, consumption, and trade in 2012. Over the next decade, aggregate ethanol production in these countries is expected to increase by 40 %. Biodiesel production is expected to rise 30 %. In the EU, corn and wheat feedstocks are expected to account for more than 80 % of the expansion of corn ethanol. Ethanol production in Brazil is also expected to increase substantially. The USDA projects that sugarcane ethanol production within the country will increase by 90 %, primarily to meet increasing domestic demand for transportation fuel with higher ethanol blends. However, exports to the EU and USA are also expected to increase. Corn-based ethanol production is expected to double in Argentina by 2022. In Canada, ethanol production is expected to increase by 35 %, with corn imports accounting for an increasing share of the feedstock. According to the USDA, China used 4.6 million tons of corn and 1 million tons of wheat to produce ethanol in 2012. Due to policies to limit the expansion of grain- and oilseed-based biofuel production, no significant expansion is expected.

Regarding US corn production, the USDA projects corn acreage will remain high in the near term, with normal yields leading to an increase in production and recovery of corn use. According to the USDA, after several years of adjusting markets, increasing producer returns are expected to lead to gradually increasing corn acreage in a range of 88 million–92 million after 2015. While projected increases in corn-based ethanol production are expected to be much smaller over the next decade, ethanol will remain a strong presence in the sector. The report notes that approximately 35 % of total corn use is expected to go to ethanol production through 2022.

More than 30 countries have introduced or are interested in introducing programs for fuel ethanol (Rosillo-Calle and Walter 2006). Other countries, but to a lower extent, have done the same regarding biodiesel. Thus, the ethanol experience is so far much more important than with biodiesel, excluding Europe where the prospects for biodiesel use are much better than fuel ethanol due to the availability of the feedstock.

Developing countries have a reasonable good potential for biofuels production due to the availability of land, better weather conditions, and the availability of cheaper labor force. Other important issue to be taken into account is that strengthening rural economies is an imperative task for these countries. Obviously, each country is case specific and a careful analysis is required to assess the pros and cons of large-scale biofuels production, particularly competition for land and water for food production and potential pressures on food prices (Hazell and von Braun 2006; Hazell and Pachauri 2006).

From an environmental perspective, first it should be highlighted the benefits of phasing out lead from gasoline, as lead has adverse neurological effects. Hydrated ethanol has higher octane number than regular gasoline (Joseph 2005), and its use in blends allows the phasing out of lead at low cost. This would be a very important advantage of ethanol use in countries where lead is still in use, as is the case of many African and some Asian and Latin American countries.

Changing over toward clean and renewable fuel from crude oil-based fuels is the need of developing countries to protect the environment. Large-scale use of biofuels is one of the main strategies for the reduction in GHG emissions (IPCC 2001). Despite the fact developing countries currently do not have biding GHG reduction targets under the Kyoto Protocol, there are two main aspects to be considered. First, under the Clean Development Mechanism (CDM), developing countries can sell credits to those with reduction commitments, as far biofuel programs are not business as usual. Considering a typical Brazilian figure of 2.7 kg of CO_2 equivalent avoided per liter of anhydrous ethanol, biofuels use could represent additional income of US\$0.02–0.05 per liter (on credits in the range US\$7-20 per tonne of CO_2 equivalent), value that should be compared with production costs in the US\$0.23–0.28 per liter range (Nastari 2005a,b). Second, climate change effects are supposed to be worst in developing countries and it is important to start acting.

Full development of the international market in fuel ethanol will require the diversification of production, in terms of both feedstocks and number of producing countries, technological development in the manufacturing field, favorable policies to induce market competitiveness and sustainable development. Bioethanol production based on lignocellulosic biomass is the technology of the future. Lignocellulosic ethanol is made from a wide variety of plant materials, including wood wastes, crop residues, and grasses, some of which can be grown on marginal lands not suitable for food production (Ghosh and Ghose 2003). Lignocellulosic raw materials minimize the potential conflict between land use for food (and feed) production and energy feedstock production. The raw material is less expensive than conventional agricultural feedstock and can be produced with lower input of fertilizers, pesticides, and energy. Biofuels from lignocellulose generate low net greenhouse gas emissions, reducing environmental impacts, particularly climate change (Hägerdal et al. 2006).

The GHG balance of biofuels varies dramatically depending on such factors as feedstock choice, associated land use changes, feedstock production system, and

Table 1.6   Estimated GHG reductions for different feedstock

Feedstock	GHG reduction (%)
Fibers (switchgrass, poplar)	70–110
Wastes (waste oil, harvest residues, sewage)	65–100
Sugars (sugarcane, sugar beet)	40–90
Vegetable oils (rapeseed, sunflower seed, soybeans)	45–75
Starches (corn, wheat)	15–40

Based on www.worldwatch.org/biofuels-transportation-selected-trends-and-facts

the type of processing energy used. In general, most currently produced biofuels have a solidly positive GHG balance. The greatest GHG benefits will be achieved with cellulosic inputs as mentioned above. Energy crops have the potential to reduce GHG emissions by more than 100 % (relative to petroleum fuels) because such crops can also sequester carbon in the soil as they grow. The estimated GHG reductions for different feedstock are shown in Table 1.6.

References

ANP (2007) Agência Nacional do Petróleo, Gás Natural e Biocombustíveis

Badger PC (2002) Ethanol from cellulose: a general review. In: Janick J, Whipkey A (eds) Trends in new crops and new uses. ASHS Press, Alexandria VA USA

Bajpai P (2007) Bioethanol. PIRA Technology Report, Smithers PIRA, UK

Berg C (2004) World Ethanol production. The distillery and bioethanol network. Available at http://www.distill.com/ world ethanol production.htm

BP (2006) The global ethanol industry is going through a period of rapid growth. Available at: http://www.bp.com/

Campbell CJ, Laherrere JH (1998) The end of cheap oil. Scientif Amer 278(3):60–65

Chandel AK, Chan ES, Rudravaram R, Lakshmi, M, Venkateswar Rao and Ravindra, P, (2007) Economics and environmental impact of bioethanol production technologies: an appraisal, Biotechnol. Mole Boil Rev 2(1): 014–032

Dien BS, Jung HJG, Vogel KP, Casler MD, Lamb JAFS, Iten L, Mitchell RB, Sarath G (2006) Chemical composition and response to dilute acid pretreatment and enzymatic saccharification of alfalfa, reed canary grass and switch grass. Biomass Bioenergy 30(10):880–891

DOE, Department of Energy (2007a) Office of Energy Efficiency and Renewable Energy. Washington DC, US, http://www.doe.gov

DOE (2007b) Fuel cell overview. US Department of Energy; available at http://hydrogen.energy.gov/

EBIO—European Biofuel Association (2006) Bioethanol fuel in numbers. Available at: www.ebio.org

EIA (Energy Information Administration) (1999) Biofuel: better for the environment United States

EIA (Energy Information Administration) (2006) Annual energy outlook 2007. US Department of Energy. Available at www.eia.doe.gov/oiaf/ieo/index.html

Ghosh P, Ghose TK (2003) Bioethanol in India: recent past and immediate future. Advances in Biochemical Engineering/Biotechnology, vol 85, p 1

Goldemberg J, Monaco LC, Macedo IC (1993) The Brazilian Fuel-Alcohol program. In: Johansson T, Kelly H, Reddy AKN, Williams RH, Burnham L (eds) Renewable energy. Sources for fuels and electricity. Island Press, Washington, pp 841–863

Gray KA, Zhao L, Emptage M (2006) Bioethanol. Curr Opin Chem Biol 10:141–146

Hägerdal BH, Galbe M, Grauslund MFG, Lidén G, Zacchi G (2006) Bioethanol: the fuel of tomorrow from the residues of today. Trends Biotechnol 24:549–556

Hamelinck CN, Hooijdonk GV, Faaij APC (2005) Ethanol from lignocellulosic biomass. Biomass Bioenergy 28:384–410

Hazell P, Pachauri RK (2006) Bioenergy and agriculture: promises and challenges. International Food Policy Research Institute

Hazell P, von Braun J (2006) Biofuels: a win–win approach that can serve the poor. International Food Policy Research Institute

Herrera S (2004) Industrial biotechnology—a chance at redemption. Nature Biotechnol 22:671–675

IPCC (2001) Intergovernmental panel on climate change. Climate change: the scientific basis. Cambridge University Press

Jessel Al (2006) Chevron Products Company 2006 Management Briefing Seminars Traverse City, MI www.cargroup.org/mbs2006/documents/JESSEL.pdf

Joseph Jr H (2005) Long term experience from dedicated & flex fuel ethanol vehicles in Brazil. Clean Vehicles and Fuels Symposium. Stockholm

Kheshgi HS, Prince RC, Marland G (2000) The potential of biomass fuels in the context of global climate change. Focus on transportation fuels. Annual Rev Energy Environ 25:199–244

Launder, Kelly (1999) Opportunities and Constraints for Ethanol-Based Transportation Fuels. Lansing: State of Michigan, Department of Consumer & Industry Services, Biomass Energy Program. Available at: http://www.michigan.gov/cis/0,1607,7-154-25676_25753_30083-141676--,00.html

Luhnow D, Samor G (2006) As Brazil fills up on Ethanol, It weans off, energy imports. Wall Street J

Lynch DJ (2006) Brazil hopes to build on its Ethanol success. USA Today

Lynd LR (1996) Overview and evaluation of fuel ethanol from cellulosic biomass: technology, economics, the environment, and policy, Annual Reviews. Energy Environ 21:403–465

Lynd LR, Wang MQ (2004) A product-nonspecific frame work for evaluating the potential of biomass-based products to displace fossil fuels. J Ind Ecol 7:17–32

McMillan JD (1997) Bioethanol production: status and prospects. Renewable Energy 10:295–302

Moreira JR, Walter A (2005) Overview on bioenergy activity for transport in Brazil. Presentation at 14th European Biomass Conference and Exhibition. Paris

Morris D (1993) Ethanol: a 150 year struggle toward a renewable future. Washington: Institute for Local Self-Reliance. Available at: www.eere.energy.gov/afdc/pdfs/1854.pdf

Nastari P (2005a) Etanol de Cana-de-Açúcar: o Combustível de Hoje. Presentation at Proalcool—30 anos depois. São Paulo

Nastari P, Macedo IC, Szwarc A (2005b) Observations on the Draft Document entitled "Potential for biofuels for transport in developing countries". Presented at the World Bank. Washington

Paszner L (2006) Bioethanol: fuel of the future. Pulp Pap Can 107(4): 26–27, 29

REN21 (2006) Renewables—Global Status Report. Renewable Energy Policy Network for the 21st Century. Available at: www.ren21.net

RFA—Renewable Fuels Association (2006) Ethanol industry outlook: From niche to nation. Available at: www.ethanolrfa.org/objects/pdf/outlook/outlook_2006.pdf

Rosillo-Calle F, Walter A (2006) Global market for bioethanol: historical trends and future prospects. Energy Sustain Dev X(1):18–30

Sun Y, Cheng J (2004) Hydrolysis of lignocellulosic materials for ethanol production: a review. Biores Technol 83:1–11

Tanaka L (2006) Ethanol fermentation from biomass resources: current state and prospects. Appl Microbiol Biotechnol 69:627–642

Turkenburg WC (2000) Renewable energy technologies (Chapter 7). In: Goldemberg J et al. (eds.) World energy assessment report, United Nations Development Programme UNDP, New York, NY, USA. pp 135–171

Walker GM (2010) Bioethanol: Science and technology of fuel alcohol, Ventus Publishing ApS ISBN 978-87-7681-681-0

Wyman CE, Hinman ND (1990). Ethanol. Fundamentals of production from renewable feedstocks and use as transportation fuel. Appl Biochem Biotechnol 24/25:735–775

Yacobucci B, Womach J (2003) Fuel Ethanol: background and public policy issues. Washington DC: Library of Congress. Available at: http://www.ethanol-gec.org/information/briefing/1.pdf

Chapter 2
Chemistry, Types, and Sources of Ethanol

2.1 Chemistry of Ethanol

Ethanol is a clear colorless, volatile, and flammable liquid that is made by the fermentation of different biological materials. Ethanol is also called ethyl alcohol or grain alcohol. It has a characteristic, agreeable odor. In dilute aqueous solution, it has a somewhat sweet flavor, but in more concentrated solutions, it has a burning taste. Ethanol is an alcohol, a group of chemical compounds whose molecules contain a hydroxyl group, –OH, bonded to a carbon atom. It may be shown as:

$$
\begin{array}{c}
\text{H} \quad \text{H} \\
| \quad | \\
\text{H-C-C-O-H} \qquad \text{or} \qquad \text{CH}_3\text{CH}_2\text{OH} \\
| \quad | \\
\text{H} \quad \text{H}
\end{array}
$$

The word alcohol derives from Arabic al-kuhul, which denotes a fine powder of antimony used as an eye makeup. Alcohol originally referred to any fine powder, but medieval alchemists later applied the term to the refined products of distillation, and this led to the current usage. Ethanol melts at −114.1 °C, boils at 78.5 °C, and has a density of 0.789 g/mL at 20 °C. Its low freezing point has made it useful as the fluid in thermometers for temperatures below −40 °C, the freezing point of mercury, and for other low-temperature purposes, such as for antifreeze in automobile radiators (Table 2.1). The molecular weight is 46.07. One gallon of 190 proof ethanol weighs 6.8 pounds. Ethanol has no basic or acidic properties. When burned, ethanol produces a pale blue flame with no residue and considerable energy, making it an ideal fuel. Ethanol mixes readily with water and with

---

Some excerpts taken from Bajpai (2007). PIRA Technology Report on Bioethanol with kind permission from Smithers PIRA

Table 2.1 Physicochemical properties of ethanol

Molecular formula	C_2H_5OH
Molecular mass	46.07 g/mol
Appearance	Colorless liquid (between -117 and 78 °C)
Water solubility	Miscible
Density	0.789 kg/l
Boiling temperature	78.5 °C (173 °F)
Freezing point	-117 °C
Flash point	12.8 °C (lowest temperature of ignition)
Ignition temperature	425 °C
Explosion limits	Lower 3.5 % v/v; upper 19 % v/v
Vapor pressure	@38 °C 50 mmHg
Higher heating value (at 20 °C)	29,800 kJ/kg
Lower heating value (at 20 °C)	21,090 kJ/L
Specific heat, Kcal/Kg	60 °C
Acidity (pKa)	15.9
Viscosity	1.200 mPa·s (20 °C)
Refractive index (nD)	1.36 (25 °C)
Octane number	99
Carbon (wt)	52.1 %
Hydrogen (wt)	13.1 %
Oxygen (wt)	34.7 %
C/H ratio	4

Based on Walker (2010)

most organic solvents. It is also useful as a solvent and as an ingredient when making many other substances including perfumes, paints, lacquer, and explosives. The *flash point* of ethanol is the lowest temperature (i.e., 12.8 °C) where enough fluid can evaporate to form an ignitable concentration of vapor and characterizes the temperature at which ethanol becomes flammable in air. The ignition point of ethanol is the minimum temperature at which it is able to burn independently (i.e., 425 °C). Ethanol has a high octane rating (99), which is a measure of a fuel's resistance to preignition, meaning that internal combustion engines using ethanol can have a high compression ratio giving a higher power output per cycle. Regular petrol (gasoline) has an average octane rating of 88. Ethanol's higher octane rating increases resistance to engine knocking, but vehicles running on pure ethanol have fuel consumption (miles per gallon or kilometers per liter) 10–20 % less than petrol (but with no loss in engine performance/acceleration).

Ethanol has been made since ancient times by the fermentation of sugars. All beverage ethanol and more than half of industrial ethanol is still made by this process. Simple sugars are the raw material. Zymase, an enzyme from yeast, changes the simple sugars into ethanol and carbon dioxide. The fermentation reaction represented by the simple equation

$$C_6H_{12}O_6 \longrightarrow 2\ CH_3CH_2OH\ +\ 2\ CO_2$$

It is actually very complex and impure cultures of yeast produce varying amounts of other substances, including glycerin and various organic acids. In the production of beverages, such as whiskey and brandy, the impurities supply the flavor. Starches from potatoes, corn, wheat, and other plants can also be used in the production of ethanol by fermentation. However, the starches must first be broken down into simple sugars. An enzyme released by germinating barley, diastase, converts starches into sugars. Thus, the germination of barley, called malting, is the first step in brewing beer from starchy plants, such as corn and wheat.

2.2 Types of Ethanol

Ethanol can be produced in two forms—hydrous and anhydrous. Hydrous ethanol is usually produced by distillation from biomass fermentation, and it contains some water residue. It is suitable for use as neat spark ignition fuel in warm climates such as that in Brazil. A further process of dehydration is required to produce anhydrous ethanol (100 % ethanol) for blending with petrol. Anhydrous ethanol can be used as an automotive fuel by itself or can be mixed with petrol in various proportions to form a petrol/ethanol blend. Anhydrous ethanol is typically blended up to 10 percent by volume in petrol, known as E10, for use in unmodified engines. Historically, the US has supported the use of E10 blends, and more recently, Europe has adopted E10 blends. Certain materials in vehicles commonly used with petrol fuel are incompatible with alcohols, and varying degrees of modification are required depending on the percentage blend of ethanol with petrol. For this reason in the European Union (EU), all member states are required to ensure that fuel grade E5 is available in the market as a protection grade for older vehicles that are not compatible to run on E10.

2.3 Raw Materials

One of the great merits of bioethanol consists in the enormous variety of raw materials, and not only plants, from which it can be produced. The production methods vary depending on whether or not the raw material is rich in fiber. The basic materials for producing biofuels must have certain features, including high carbon and hydrogen concentrations and low concentrations of oxygen, nitrogen, and other organic components. The following is a brief description of some of the most important raw materials suitable for use in bioethanol production. These can be classified into three groups (Table 2.2):

1. Feedstocks rich in sugar
2. Starches
3. Cellulosic materials

Table 2.2 Major resources for bioethanol production

Sugary materials
Sugarcane
Sugarbeet
Sweet sorghum
Cheese whey
Fruits (surplus)
Confectionery industrial waste
Starchy materials
Grains (maize, wheat, triticale)
Root crops (potato, cassava, chicory, artichoke)
Inulin (polyfructan)
Cellulosic materials
Wood
Agricultural residues (straws, stover)
Municipal solid waste
Waste paper, paper pulp

Based on data from Hamelinck et al. (2003) and US DOE (2006a)

Agricultural waste available for ethanol conversion includes crop residues such as wheat straw, corn stover (leaves, stalks, and cobs), rice straw, and bagasse (sugarcane waste). Forestry waste includes underutilized wood and logging residues; rough, rotten, and salvable dead wood; and excess saplings and small trees (Walker 2010). Municiple solid waste contains some cellulosic materials, such as paper, energy crops, developed and grown specifically for fuel, include fast-growing trees, shrubs, and grasses such as hybrid poplars, willows, reed canary grass, alfalfa *Miscanthus x giganteus* (hybrid of *M. sinensis* and *M. sacchariflorus*), and switchgrass. Switchgrass (*Panicum virgatumn*) is one source likely to be tapped for ethanol production because of its potential for high fuel yields, hardiness, and ability to be grown in diverse areas. This is a perennial herbaceous plant that grows mainly in the United States. Its ethanol yield per hectare is the same as for wheat. It responds to nitrogen fertilizers and can sequester the carbon in the soil. It is a highly versatile plant, capable of adapting easily to lean soils and marginal farmland (Heaton et al. 2004). Like maize, it is a type C4 plant, i.e., it makes an alternative use of CO_2 fixation (a process forming part of photosynthesis). Most of the genotypes of Panicum virgatum have short underground stems, or rhizomes, that enable them with time to form a grassy carpet. Single hybrids of Panicum virgatum have shown a marked potential for increasing their energy yield (Bouton 2007), but genetic engineering methods on this plant are still in a developmental stage and for the time being only their tetraploid and octaploid forms are known; we also now know that similar cell types (isotypes) reproduce easily.

Trials show current average yields to be about five dry tons per acre; however, crop experts say that progressively applied breeding techniques could more than double that yield. Switchgrass long root system—actually a fifty–fifty split above ground and below—helps keep carbon in the ground, improving soil quality. It is drought-tolerant, grows well even on marginal land, and does not require heavy fertilizing. Other varieties including big blue stem and Indian grass are also possible cellulose sources for ethanol production. Researchers estimate that ethanol yield from switchgrass is in the range of 60–140 gallons per ton; some say 80 to 90 gallons per ton is a typical figure. It is estimated that the energy output/energy input ratio for fuel ethanol made from switchgrass, is about 4.4 (Iowa State University 2006). The US Department of Agriculture estimates that by 2030 approximately 129 million acres of excess cropland could be used for energy crops. If 40 million of these acres were utilized for energy crops for biofuels such as ethanol, it would provide a transportation fuel equivalent to 550 million barrels of oil per year. Sugarcane bagasse, the residue generated during the milling process, is another potential feedstock for cellulosic ethanol. Research shows that one ton of sugarcane bagasse can generate 112 gallons of ethanol.

Reed canary grass (*Phalaris arundinacea L*) is a perennial grass that grows widely. Its stem components (dry wt) comprise: hexoses (38–45 %); pentoses (22–25 %); lignin (18–21 %). Alfalfa (*Medicago sativa L*) comprises mainly cellulose, hemicellulose, lignin, pectin, and protein. *Miscanthus x giganteus* (hybrid of *M. sinensis* and *M. sacchariflorus*) is a perennial grass with a low need for fertilzers and pesticides with a broad temperature growth range. Previously used as an ornamental landscaping, but now an attractive biomass source for biofuels. For example, potential ethanol from miscanthus is around twice that from corn on an acreage basis.

The most suitable waste for converting into bioethanol is the waste from the fruit and vegetable industries, for instance, cotton fiber, milk whey from cheese-making, the waste products of coffee making, and so on. Generally speaking, such waste contains approximately 45 % of cellulose (glucose polymer), which can be simultaneously hydrolyzed and fermented to produce ethanol. SSL (spent sulfite liquor) is a byproduct of bisulfate "pulp" manufacturing that can also be fermented to produce ethanol. Waste varies considerably in content from one area to another, but the majority of the volume generally consists of paper (20–40 %), gardening waste (10–20 %), plastics, glass, metals, and various other materials (Prasad et al. 2007).

Jerusalem artichoke (*Helianthus tuberosus*) tubers are rich in inulin (a fructose polymer), which can be used to obtain a syrup for use both in the foodstuffs industry and in the production of ethanol. It was demonstrated (Curt et al. 2006) that, toward the end of the season, the potential for bioethanol production of the stems of clones is 38 % of that of the tubers. The plant grows in summer, reaching its maximum height in July and dying in October.

There are certain other feedstocks which can be used for bioethanol production. Marine macroalgae (seaweeds) demand minimal use of agriculture areas and fresh water for cultivation and represent an interesting biomass resource for bioethanol (Horn et al. 2000a, b). Their attraction as biomass sources for biofuels stems from the fact that seaweeds have growth rates and primary production rates far in excess than those of terrestrial plants. As low-input, high-yielding biomass, seaweeds may represent an example of *third generation* feedstocks for bioethanol production. Brown seaweeds (Phaeophyta) in particular contain storage polysaccharides which are substrates for microbial degradation. They contain high amounts of carbohydrates such as alginic acid (structural) and laminaran and mannitol (storage) that can potentially be fermented to ethanol. Alginate typically makes up 30–40 % of the dry weight in giant brown seaweeds (kelp). Laminarin is a linear polysaccharide of 1,3-β-Dglucopyanose and can relatively easily be hydrolyzed to fermentable glucose. Unlike lignocellulosic biomass, they have low levels of lignin and cellulose making them more amenable for bioconversion to energy fuels than terrestrial plants. Fermentations of hydroyzates derived from the fast-growing *Macrocystis* spp and *Laminaria* spp. hold the greatest potential for marine macroalgal bioethanol. Another substance present in great quantities in the sea is chitin which is a polysaccharide consisting of N acetyl glucosamine monomers. Chitin is a very hard, semi-transparent substance found naturally in the exoskeletons of crabs, lobsters, and shrimps. Its structure resembles that of cellulose except one hydroxyl group is replaced by an acetyl amine group. It has been described as "cellulose of the sea" and has potential for bioconversion into chemical commodities, including ethanol. A major coproduct of biodiesel production is glycerol, which has the potential to be converted to ethanol by certain bacteria (Dharmadi et al. 2006) and yeasts such as *Candida magnoliae, Zygosaccharomyces rouxii,* and *Pachysolen tannophilus.*

Lignocellulose consists of three main components: cellulose, hemicellulose, and lignin, the first two being composed of chains of sugar molecules. These chains can be hydrolyzed to produce monomeric sugars, some of which can be fermented using ordinary baker's yeast. To attain economical feasibility, a high ethanol yield is a necessity. However, producing monomer sugars from cellulose and hemicellulose at high yields is far more difficult than deriving sugars from sugar- or starch-containing crops, e.g., sugarcane or maize. Therefore, although the cost of lignocellulosic biomass is far lower than that of sugar and starch crops, the cost of obtaining sugars from such materials for fermentation into bioethanol has historically been far too high to attract industrial interest. For this reason, it is crucial to solve the problems involved in the conversion of lignocellulose to sugar and further to ethanol. However, the heterogeneity in feedstock and the influence of different process conditions on microorganisms and enzymes makes the biomass-to-ethanol process complex. Table 2.3 shows research challenges in the field of bioethanol production based on lignocellulosic biomass.

Table 2.3 Research challenges in the field of bioethanol production based on lignocellulosic biomass

Major challenges
Improving the enzymatic hydrolysis with efficient enzymes, reduced enzyme production cost and novel technology for high solids handling
Developing robust fermenting organisms, which are more tolerant to inhibitors and ferment all sugars in the raw material in concentrated hydrolyzates at high productivity and with high concentration of ethanol
Extending process integration to reduce the number of process steps and the energy demand and to reuse process streams for eliminating use of fresh water and to reduce the amount of waste streams

Based on Hahn et al. (2006)

Table 2.4 Biochemical composition of some Lignocellulosic feedstock (% dry basis)

Feedstock	Hardwood (eucalyptus)	Softwood (pine)	Grass (switchgrass)
Cellulose	49.50	44.55	31.98
Hemicellulose	13.07	21.90	25.19
Lignin	27.71	27.67	18.13
Ash	1.26	0.32	5.95
Acids	4.19	2.67	1.21
Extractives	4.27	2.88	17.54
Heating values ($GJ_{HHV}/tonne_{dry}$)	19.5	19.6	18.6

Sun and Cheng (2002) and Mosier et al. (2005)

After the cellulose and hemicellulose have been saccharified, the remainder of the ethanol production process is similar to grain-ethanol. However, the different sugars require different enzymes for fermentation.

Lignocellulosic crops are promising feedstock for ethanol production because of high yields, low costs, good suitability for low quality land and low environmental impact. Most ethanol conversion systems encountered in literature have been based on a single feedstock. But considering the hydrolysis fermentation process, it is possible to use multiple feedstock types. Table 2.4 presents biochemical compositions for several suitable feedstock—eucalyptus, Pinus, and switchgrass. Pine has the highest combined sugar content, implying the highest potential ethanol production. The lignin content for most feedstock is about 27 % but grasses contain significantly less and will probably coproduce less electricity.

Cellulosic resources are in general very widespread and abundant. For example, forests comprise about 80 % of the world's biomas. Being abundant and outside the human food chain makes cellulosic materials relatively inexpensive feedstocks for ethanol production. Brazil uses sugarcane as primary feedstock whereas in US more than 90 % of the ethanol produced comes from corn. Other feedstocks such as beverage waste, brewery waste, and cheese whey are also being utilized. In European union, most of the ethanol is produced from sugar beets and wheat. Crops with higher yields of energy, such as switchgrass and sugarcane, are more effective in producing ethanol than corn. Ethanol can also be produced from sweet sorghum, a dryland crop that uses much less water than sugarcane, does not require a tropical climate, and produces food and fodder in addition to fuel. The best farm crop for ethanol production is sugar beets, in terms of gallons of fuel per acre, with the lowest water requirements to grow the crop (The beet plant drives a central taproot deep into the soil and the entire beet is underground, minimimizing evaporation). One result of increased use of ethanol is increased demand for the feedstocks. Large-scale production of agricultural alcohol may require substantial amounts of cultivable land with fertile soils and water. This may lead to environmental damage such as deforestation or decline of soil fertility due to reduction of organic matter. About 5 % (in 2003) of the ethanol produced in the world is actually a petroleum product. It is made by the catalytic hydration of ethylene with sulfuric acid as the catalyst. It can also be obtained via ethylene or acetylene, from calcium carbide, coal, oil gas, and other sources. Two million tons of petroleum-derived ethanol are produced annually. The principal suppliers are plants in the United States, Europe, and South Africa. Petroleum-derived ethanol (synthetic ethanol) is chemically identical to bioethanol and can be differentiated only by radiocarbon dating.

References

Bajpai P (2007) Bioethanol. PIRA Technology Report, Smithers PIRA, UK

Bouton JH (2007) Molecular breeding of switchgrass for use as a biofuel crop. Curr Opin Genet Dev 17:553–558

Curt MD, Aguado P, Sanz M, Sánchez G, Fernández J (2006) Clone precocity and the use of *Helianthus tuberosus* L. stems for bioethanol. Ind Crops Prod 24(3):314–320

Department of Energy (2006a) Bioethanol feedstocks. Available at: http://www1.eere.energy.gov/biomass/abcs_biofuels.html

Dharmadi Y, Muraka A, Gonzalez R (2006) Anerobic fermentation of glycerol by *Escherichia coli*: a new platform for metabolic engineering. Biotechnol Bioeng 94:821

Hahn Hagerdal B, Galbe M, Gorwa-Grauslund MF, Liden G, Zacchi G (2006) Bioethanol—the fuel of tomorrow from the residues of today. *Trends Biotechnol* 24:549–556

Hamelinck CN, Hooijdonk GV, Faaij APC (2003) Prospects for ethanol from lignocellulosic biomass: techno-economic performance as development progresses. Universiteit Utrecht Copernicus Institute, Science Technology Society, NWS-E-2003-55 ISBN 90-393-2583-4, Available at: www.senternovem.nl/mmfiles/149043_tcm24-124362.pdf

Heaton E, Voigt T, Long SP (2004) A quantitative review comparing the yields of two candidate C4 perennial biomass crops in relation to N, temperature and water. Biomass Bioenergy 27:21–30

Horn SJ, Aasen IM, Østgaard K (2000a) Production of ethanol from mannitol by *Zymobacter palmae*. J Ind Microbiol Biotechnol 24:51–57

Horn SJ, Aasen IM, Østgaard K (2000b) Ethanol production from seaweed extract. J Ind Microbiol Biotechnol 25:249–254

Iowa State University (2006). Biomass economics. http://www.public.iastate.edu/~brummer/ag/biomass2.htm

Mosier N, Wyman C, Dale B, Elander R, Lee YY, Holtzapple M, ladisch M (2005) Features of promising technologies for pretreatment of lignocellulosic biomass. Bioresour Technol 96:673–686

Prasad S, Singh A, Joshi HC (2007) Ethanol as an alternative fuel from agricultural, industrial and urban residues. Resour Conserv Recycl 50(1):1–39

Sun Y, Cheng J (2002) Hydrolysis of lignocellulosic materials for ethanol production: a review. Bioresour Technol 83:1–11

Walker GM (2010). Bioethanol: Science and technology of fuel alcohol. Ventus Publishing, London. ApS ISBN 978-87-7681-681-0

Chapter 3
Production of Bioethanol

Bioethanol production processes vary considerably depending on the raw material involved, but some of the main stages in the process remain the same, even though they take place in different conditions of temperature and pressure, and they sometimes involve different microorganisms. These stages include hydrolysis (achieved chemically and enzymatically), fermentation and distillation (Olsson et al. 2005). Today, there are mainly two types of process technology called first- and second-generation technology (Fig. 3.1).

First-generation process technology
First-generation process technology produces bioethanol from sugars (a dimer of the monosaccharides glucose and fructose) and starch-rich (polysaccharides of glucose) crops such as grain and corn (Table 3.1). Sugars can be converted to ethanol directly but starches must first be hydrolyzed to fermentable sugars by the action of enzymes from malt or molds. The technology is well-known but high prices of the raw material and the ethics about using food products for fuel are two major problems.

Second-generation process technology
The raw material in second generation is lignocellulosic materials such as straw, wood, and agricultural residues, which are often available as wastes (Table 3.1). These kinds of materials are cheap but the process technology is more advanced than converting sugar and starch (Fan et al. 1987; Badger 2002). Basically, the lignocellulosic biomass comprises lignin, cellulose, and hemicelluloses (Figs. 3.2, 3.3a–c). Cellulose is a linear, crystalline homopolymer with a repeating unit of glucose strung together beta-glucosidic linkages. The structure is rigid, and harsh treatment is required to break it down (Gray et al. 2006). Hemicellulose consists of short, linear, and highly branched chains of sugars. In contrast to cellulose, which is a polymer of only glucose, a hemicellulose is a hetero-polymer of D-xylose, D-glucose, D-galactose, D-mannose, and L-arabinose. The composition of holocellulose (cellulose + hemicellulose) varies with the origin of the lignocellulosic material.

Some excerpts taken from Bajpai (2007). PIRA Technology Report on Bioethanol with kind permission from Smithers PIRA

P. Bajpai, *Advances in Bioethanol*, SpringerBriefs in Applied Sciences and Technology,   21
DOI: 10.1007/978-81-322-1584-4_3, © The Author(s) 2013

Table 3.1 First-generation and second-generation feedstocks for bioethanol

First-generation feedstocks
Sugar beet
Sweet sorghum
Sugarcane
Maize
Wheat
Barley
Rye
Grain
Sorghum
Triticale
Cassava
Potato
Second-generation feedstocks
Corn stover
Wheat straw
Sugarcane bagasse
Municipal solid waste

Based on Walker (2010)

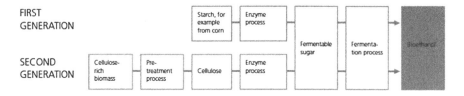

Fig. 3.1 Bioethanol production processes. Based on Novozymes report 2007

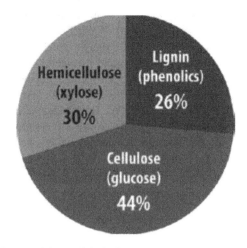

Fig. 3.2 Main constituents of lignocellulosic feedstocks

(a)

(b)

(c)

Non-Reducing
End-Group

Anhydroglucose unit, AGU
(n = value of DP)

Reducing
End-Group

Fig. 3.3 **a** Structure of lignin [complex crosslinked polymer of aromatic rings (phenolic mono-mers); very high energy content]. **b** Structure of hemicellulose (branching polymers of C5, C6, uronic acid, acetyl derivatives). **c** Structure of cellulose (composed of D-glucose unite linked by β-1, 4 glycoside bonds). Based on Walker (2010)

Burning ethanol obtained from cellulose produces 87 % lower emissions than burning petrol, while for the ethanol from cereals, the figure is no more than 28 %. Ethanol obtained from cellulose contains 16 times the energy needed to produce it, petrol only 5 times, and ethanol from maize only 1.3 times. The problem is a matter of how to disrupt the bonds of this molecule in order to convert it into fermentable sugars. In fact, this is unquestionably the type of raw material that is the most complicated to process. Lignin binds together pectin, protein and the two types of polysaccharides, cellulose and hemicellulose, in lignocellulosic biomass. Lignin resists microbial attack and adds strength to the plant. Pretreatment is therefore used to open the biomass by degrading the lignocellulosic structure and releasing the polysaccharides (Fig. 3.4) (Mosier et al. 2005; Ladisch 2003; Hsu et al. 1980). Pretreatment is followed by treatment with enzymes, which hydrolyze cellulose and hemicellulose, respectively. The cellulose fraction releases glucose (C6 monosaccharide—sugar with six carbon atoms), and the hemicellulose fraction releases pentoses (C5 monosaccharide—sugar with five carbon atoms) such as xylose. Out of carbohydrate monomers in lignocellulosic materials, xylose is second most abundant after glucose. Glucose is easily fermented into ethanol, but another fermentation process is required for xylose—for example using special microorganisms. The second generation holds great advantages with the fermentation of biomass in form of agricultural waste materials, but there are some challenges such as efficient pretreatment and fermentation technologies together with environmentally friendly process technology (for example reuse of the process water). Figure 3.5 shows schematic of a biochemical, and Fig. 3.6 shows schematic of thermochemical cellulosic ethanol production process.

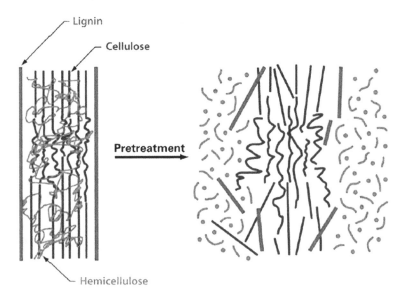

Fig. 3.4 Pretreatment to open the biomass by degrading the lignocellulosic structure and releasing the polysaccharides. Based on Mosier et al. (2005), Ladisch (2003), and Hsu et al. (1980)

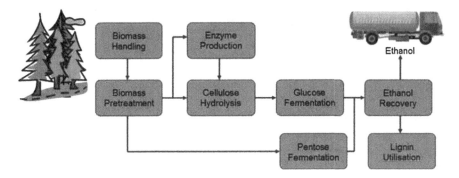

Fig. 3.5 Schematic of a biochemical cellulosic ethanol production process. Johnson et al. (2010), reproduced with permission

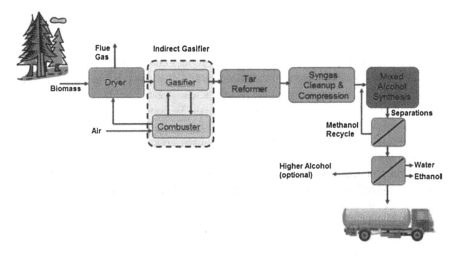

Fig. 3.6 Schematic of a cellulosic ethanol production process through gasification technology. Johnson et al. (2010), reproduced with permission

3.1 Feedstock Processing

Processing bioethanol feedstocks include preprocessing, pretreatment; hydrolysis, and microbial contamination control. The salient features for processing sugarcane, maize, and lignocellulose feedstocks are presented below.

3.1.1 Processing of Sugar Crops

Sugarcane contains ~15 % sucrose. The juice is pressed from the cane and is fermented by yeast. The juice can be processed either into crystalline sugar or directly

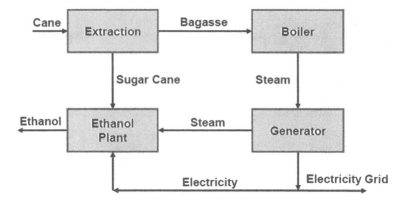

Fig. 3.7 Processing of sugarcane to ethanol. Based on Walker (2010)

fermented to ethanol, as per many industrial plants in Brazil (Fig. 3.7). For sugar production, the juice is clarified with lime and evaporated to form crystals that are centrifuged, leaving a syrupy brown liquid by-product known as molasses. Molasses represents an almost complete fermentation medium as it comprises sugars (sucrose, glucose, fructose), minerals, vitamins, fatty acids, organic acids, etc. Additional nitrogen in the form of diammonium phosphate is commonly added. The more sucrose from sugarcane stalks that is removed for crystalline sugar production, the poorer the quality of molasses and some molasses contains excess levels of salts and inhibitors produced during heat treatments (furfurals, formic acid, and browning reaction products). For bioethanol fermentations, molasses is diluted to 20–25 % total sugar treated with sulfuric acid and heated to 90 °C for impurity removal prior to cooling, centrifugation, pH adjustment, and addition of yeast. Sugarcane juice can either be directly fermented, clarified following heat (105 °C) treatment, or mixed with molasses in different proportions. Constituents in molasses that are important for bioethanol production include sugar content: sugar % (w/w) and degrees Brix, color, total solids, specific gravity, crude protein, free amino nitrogen, total fat, fiber, minerals, vitamins, and substances toxic to yeast. The yeast *S. cerevisiae* is the predominant microorganism employed in industrial molasses fermentations, but another yeast, *Kluyveromyces marxianus*, and a bacterium, *Zymomonas mobilis*, have potential in this regard (Senthilkumar and Gunasekaran 2008).

3.1.2 Processing of Cereal Crops

The main stages prior to fermentation for processing of starch-based materials are Cereal cooking; Starch liquefaction, and Amylolysis. In North America, ethanol is produced from corn by using one of two standard processes: wet milling or dry milling (Yacobucci and Womach 2003; Ferguson 2003) (Figs. 3.8, 3.9). The main difference between the two is in the initial treatment of the grain. Dry milling plants cost less to build and produce higher yields of ethanol (2.7 gallons per bushel of corn),

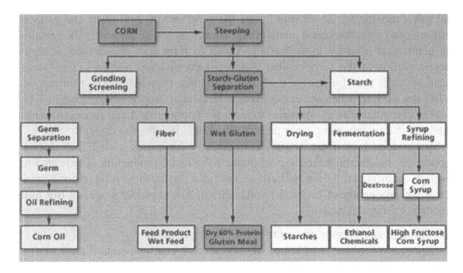

Fig. 3.8 Ethanol production process—wet milling. http://www.ethanolrfa.org/pages/how-ethanol-is-made. Reproduced with permission

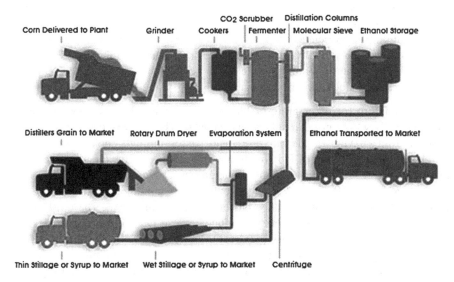

Fig. 3.9 The ethanol production process—dry milling. http://www.ethanolrfa.org/pages/how-ethanol-is-made. Reproduced with permission

but the value of the co-products is less. The value of corn as a feedstock for ethanol production is due to the large amount of carbohydrates specifically starch present in corn. In wet milling, maize kernels are soaked in water (or dilute acid) to separate the cereal into starch, gluten, protein, oil, and fiber prior to starch conversion to ethanol.

The wet-milling operation is more elaborate because the grain must be separated into its components. After milling, the corn is heated in a solution of water and sulfur dioxide for 24–48 h to loosen the germ and the hull fiber. The germ is then removed from the kernel, and corn oil is extracted from the germ. The remaining germ meal is added to the hulls and fiber to form corn gluten feed. A high-protein portion of the kernel called gluten is separated and becomes corn gluten meal, which is used for animal feed. In wet milling, only the starch is fermented, unlike dry milling, when the entire mash is fermented. In dry milling, from which most US bioethanol is made, maize kernels are finely ground and processed without fractionation into component parts (O'Brien and Woolverton 2009).

Starch-bioethanol (from US maize) currently dominates global fuel alcohol production, but the projected use of maize for ethanol production is expected to level-off (at around 6 billion bushels) unless "idle" land can be used to grow more cereal for production of biofuels (Abbas 2010). The principal stages in dry mill bioethanol processes are:

1. Milling (maize kernels ground to a fine powder or meal)
2. Liquefaction (water is added to the maize meal and temperature increased in the mash to solubilize starch)
3. Saccharification (enzymatic hydrolysis of starch liberates simple sugars, mainly glucose)
4. Fermentation (starch hydrolyzate is fermented by yeast to ethanol, CO_2, and secondary metabolites)
5. Distillation (the fermented wash, or beer, at around 10 % v/v ethanol is distilled to ~96 % v/v ethanol with the solid residues processed into animal feed)
6. Dehydration (water remaining in the ethanolic distillate is removed by molecular sieves to produce anhydrous ethanol).

After the corn (or other grain or biomass) is cleaned, it passes first through hammer mills which grind it into a fine powder. The meal is then mixed with water and an enzyme (alpha amylase), and passes through cookers where the starch is liquefied. A pH of 7 is maintained by adding sulfuric acid or sodium hydroxide. Heat is applied to enable liquefaction. Cookers with a high temperature stage (120–150 °C) and a lower temperature holding period (95 °C) are used. The high temperatures reduce bacteria levels in the mash. The mash from the cookers is cooled, and the enzyme glucoamylase is added to convert starch molecules to fermentable sugars (dextrose). Yeast is added to the mash to ferment the sugars to ethanol and carbon dioxide. Using a continuous process, the fermenting mash flows through several fermenters until the mash is fully fermented and leaves the tank. In a batch fermentation process, the mash stays in one fermenter for about 48 h. The fermented mash, now called "beer," contains about 10 % alcohol, as well as all the non-fermentable solids from the corn and the yeast cells. The mash is then pumped to the continuous flow, multi-column distillation system where the alcohol is removed from the solids and water. The alcohol leaves the top of the final column at about 96 % strength, and the residue mash, called stillage, is transferred from the base of the column to the co-product processing area. The stillage is sent through a centrifuge that separates the coarse grain from the solubles. The solubles are then concentrated to about

30 % solids by evaporation, resulting in Condensed Distillers Solubles (CDS) or "syrup." The coarse grain and the syrup are then dried together to produce dried distillers grains with solubles (DDGS), a high-quality, nutritious livestock feed. The CO_2 released during fermentation is captured and sold for use in carbonating soft drinks and beverages and the manufacture of dry ice. Drying the distillers grain accounts for about 1/3 of the plants energy usage. The alcohol then passes through a dehydration system where the remaining water is removed. Most plants use a molecular sieve to capture the last bit of water in the ethanol. The alcohol at this stage is called *anhydrous* (pure, without water) ethanol and is approximately 200 proof. Ethanol that is used for fuel is then denatured with a small amount (2–5 %) of some product, like gasoline, to make it unfit for human consumption.

Starch is an alpha-polysaccharide comprising D-glucose monomers arranged in two basic formats: amylose and amylopectin (Fig. 3.10a, b), and plant starches generally contain 10–25 % amylose and 75–90 % amylopectin (depending on the biomass source). Industrial enzymes used as processing aids in starch-to-ethanol bioconversions are produced by microorganisms (bacteria such as *Bacillus*

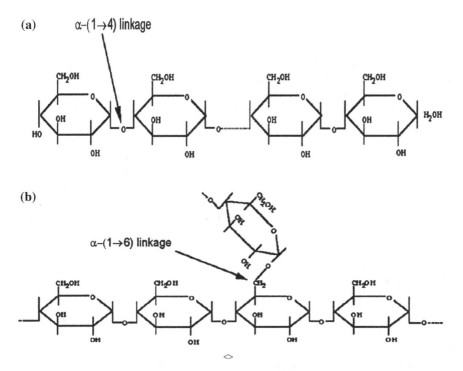

Fig. 3.10 a Structure of amylase. **b** Structure of amylopectin. Based on http://www.gtconsult.com. br/ingles/artigos/What_is_Starch.pdf, http://www.scientificpsychic.com/fitness/carbohydrates1.html

spp. and fungi such as *Aspergillus* spp.) grown in closed fermentation tanks by specialist companies (e.g., Novozymes, Genencor). The industrial production and purification of amylolytic enzymes for bioethanol production have been discussed by Nair et al. (2008). In order for starch to be converted to ethanol by yeast (*S. cerevisiae*), it has to be de-polymerized to constituent saccharides such as glucose and maltose. In traditional beverage fermentation industries such as brewing, this is partially accomplished using endogenous enzymes, mainly alpha and beta amylases, present in malted barley. However, for bioethanol production, more complete starch hydrolysis is required and this is conducted using exogenous (microbially derived) amylolytic enzymes including de-branching enzymes such as amyloglucosidase (or glucoamylase).

The production of ethanol is an example of how science, technology, agriculture, and allied industries must work in harmony to change a farm product into a fuel. Ethanol plants receive the large quantities of corn they need by truck, rail, or barge. The corn is cleaned, ground, and blown into large tanks where it is mixed into a slurry of corn meal and water. Enzymes are added, and exact acidity levels and temperatures are maintained, causing the starch in the corn to break down— first into complex sugars and then into simple sugars.

New technologies have changed the fermentation process. In the beginning, it took several days for the yeast to work in each batch. A new, faster, and less costly method of continuous fermentation has been developed. Plant scientists and geneticists are also involved. They have been successful in developing strains of yeast that can convert greater percentages of starch to ethanol. Scientists are also developing enzymes that will convert the complex sugars in biomass materials to ethanol. Cornstalks, wheat and rice straw, forestry wastes, and switchgrass all show promise as future sources of ethanol.

In modern ethanol production, for every bushel of corn that is processed, one-third is returned to the livestock feed market. That is because ethanol production requires only the starch portion of a corn kernel. The remaining protein, fat, fiber, and other nutrients are returned to the global livestock and poultry feed markets (RFA 2007a). Thus, every bushel of corn processed by an ethanol plant produces 2.8 gallons of ethanol—and approximately 17 pounds of animal feed. This high-quality feed for cattle, poultry, and pigs is not a by-product of ethanol production; it is a co-product. During 2012 alone, the U.S. ethanol industry used 4.5 billion bushels of corn to produce an estimated 13.3 billion gallons of ethanol and 34.4 million metric tons of high-quality livestock feed. This includes 31.6 million metric tons of distillers grains and 2.8 million tons of corn gluten feed and meal (Fig. 3.11) (RFA 2013).

This level of output will make it necessary to find new markets and uses for co-products. New uses being considered include food, fertilizer, and cat litter. While the majority of feed is dried and sold as Distillers Dried Grains with Solubles (DDGS), approximately 20–25 % is fed wet locally, reducing energy costs associated with drying as well as transportation costs. Ethanol wet mills produced approximately 430,000 metric tons of corn gluten meal, 2.4 million metric tons of corn gluten feed and germ meal, and 565 million pounds of corn oil. Figure 3.12 shows production of US ethanol feed co-products (RFA 2013).

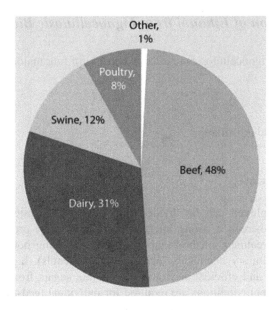

Fig. 3.11 Distillers grain consumption by species* (estimated). RFA (2013). Reproduced with permission

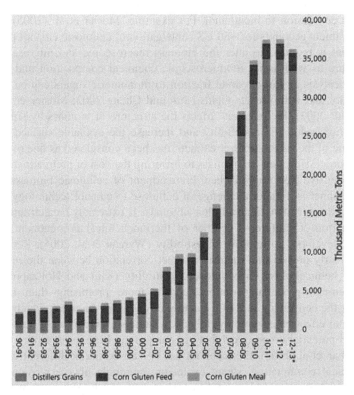

Fig. 3.12 Production of US ethanol feed co-products. RFA (2013). Reproduced with permission

3.1.3 Production of Ethanol from Lignocellulosic Biomass

Bioconversion of lignocellulosics to ethanol consists of four major unit operations:

- Pretreatment
- Hydrolysis
- Fermentation
- Product separation/distillation.

3.1.3.1 Pretreatment

Pretreatments employed can be divided into physical, chemical, and biological methods, but there is a strong inter-dependence of these processes. There is not a perfect pretreatment method employed and remaining bottlenecks include generation of inhibitory chemicals (acids, furans, phenols), high particle load, high energy input and efficient separation of soluble sugars from solid residues. Specific pretreatment conditions are required for individual feedstocks, and mechanistic models can help in the rational design of such processes (Zhang et al. 2009; Eggeman and Elander 2005). It is especially important to optimize lignocellulose pretreatment methods because they are one of the most expensive steps in the overall conversion to bioethanol. For example, Mosier et al. (2005) reported that pretreatment accounts for ~30 US cents/gallon of cellulosic ethanol produced. Pretreatment is required to alter the biomass macroscopic and microscopic size and structure as well as its submicroscopic chemical composition and structure so that hydrolysis of carbohydrate fraction to monomeric sugars can be achieved more rapidly and with greater yields (Sun and Cheng 2002; Mosier et al. 2005; Tucker et al. 2003). Pretreatment affects the structure of biomass by solubilizing hemicellulose, reducing crystallinity and increase the available surface area and pore volume of the substrate. Pretreatment has been considered as one of the most expensive processing steps in biomass to lowering the cost of pretreatment process through extensive R&D approaches. Pretreatment of cellulosic biomass in cost-effective manner is a major challenge of cellulose to ethanol technology research and development. Native lignocellulosic biomass is extremely recalcitrant to enzymatic digestion. Therefore, a number of thermochemical pretreatment methods have been developed to improve digestibility (Wyman et al. 2005). Recent studies have clearly proved that there is a direct correlation between the removal of lignin and hemicellulose on cellulose digestibility (Kim and Holtzapple 2006). Thermochemical processing options appear more promising than biological options for the conversion of lignin fraction of cellulosic biomass, which can have a detrimental effect on enzyme hydrolysis. It can also serve as a source of process energy and potential co-products that have important benefits in a life-cycle context (Sheehan et al. 2003). Pretreatment can be carried out in different ways such as mechanical combination, steam explosion, ammonia fiber explosion, acid or alkaline pretreatment and biological treatment, organosolv pretreatment (Cadoche and Lopez 1989; Gregg and Saddler 1996; Kim et al. 2003; Damaso et al. 2004; Kuhad et al. 1997; Keller et al. 2003; Itoh et al. 2003).

To asses the cost and performance of pretreatment methods, technoeconomic analysis have been made (Eggerman and Elander 2005). There is huge scope in lowering the cost of pretreatment process through extensive R&D approaches. Pretreatment of cellulosic biomass in cost-effective manner is a major challenge of cellulose to ethanol technology research and development. Native lignocellulosic biomass is extremely recalcitrant to enzymatic digestion. Therefore, a number of thermochemical pretreatment methods have been developed to improve digestibility (Wyman et al. 2005). Recent studies have clearly proved that there is a direct correlation between the removal of lignin and hemicellulose on cellulose digestibility (Kim and Holtzapple 2006). Thermochemical processing options appear more promising than biological options for the conversion of lignin fraction of cellulosic biomass, which can have a detrimental effect on enzyme hydrolysis. It can also serve as a source of process energy and potential co-products that have important benefits in a life-cycle context (Sheehan et al. 2003).

3.1.3.2 Hydrolysis

The hydrolysis methods most commonly used are acid and enzymatic. Both dilute and concentrated acid is used. Dilute acid treatment is employed for the degradation of hemicellulose leaving lignin and cellulose network in the substrate. Other treatments are alkaline hydrolysis or microbial pretreatment with white-rot fungi (*Phaenerochate chrysosporium*, *Cyathus stercoreus*, *Cythus bulleri*, and *Pycnoporous cinnabarinus*, etc.) preferably act upon lignin leaving cellulose and hemicellulose network in the residual portion. However, during both treatment processes, a considerable amount of carbohydrates are also degraded; hence, the carbohydrate recovery is not satisfactory for ethanol production. The dilute acid process is conducted under high temperature and pressure and has reaction time in the range of seconds or minutes. The concentrated acid process uses relatively mild temperatures, but at high concentration of sulfuric acid and a minimum pressure involved, which only creates by pumping the materials from vessel to vessel. Reaction times are typically much longer than for dilute acid. In dilute acid hydrolysis, the hemicellulose fraction is depolymerized at lower temperature than the cellulosic fraction. Dilute sulfuric acid is mixed with biomass to hydrolyze hemicellulose to xylose and other sugars. Dilute acid is interacted with the biomass, and the slurry is held at temperature ranging from 120–220 °C for a short period of time. Thus, hemicellulosic fraction of plant cell wall is depolymerized and will lead to the enhancement of cellulose digestibility in the residual solids (Nigam 2002; Sun and Cheng 2002; Dien et al. 2006; Saha et al. 2005). Dilute acid hydrolysis has some limitations. If higher temperatures (or longer residence time) are applied, the hemicellulosic derived monosaccharides will degrade and give rise to fermentation inhibitors like furan compounds, weak carboxylic acids, and phenolic compounds (Olsson and Hahn-Hagerdal 1996; Klinke et al. 2004;

Larsson et al. 1999). These fermentation inhibitors are known to affect the ethanol production performance of fermenting microorganisms (Chandel et al. 2007).
In order to remove the inhibitors and increase the hydrolyzate fermentability, several chemicals and biological methods have been used. These methods include
overliming (Martinez et al. 2000), charcoal adsorption (Chandel et al. 2007),
ion exchange (Nilvebrant 2001), detoxification with laccase (Martin et al. 2002;
Chandel et al. 2007), and biological detoxification (Lopez et al. 2004). The detoxification of acid hydrolyzates has been shown to improve their fermentability;
however, the cost is often higher than the benefits achieved (Palmqvist and Hahn-
Hagerdal 2000; von Sivers et al. 1994). Dilute acid hydrolysis is carried out in two
stages—First-stage and two-stage.

Enzymatic hydrolysis using cellulases does not generate inhibitors, and the
enzymes are very specific for cellulose. Cellulases, mainly derived from fungi
like *Trichoderma reesei* and bacteria like *Cellulomonas fimi*, are a mixture of at
least three different enzymes: (1) endoglucanase which attacks regions of low
crystallinity in the cellulose fiber, creating free chain-ends; (2) exoglucanase or
cellobiohydrolase which degrade the molecule further by removing cellobiose
units from the free chain-ends; and (3) β-glucosidase which hydrolyzes cellobiose to produce glucose, in much smaller amounts. Cellulase enzyme, however,
has been projected as a major cost contributor to the lignocellulose-to-ethanol
technology. The main challenges are the low glucose yield and high cost of the
hydrolysis process. Cellulase costs around 15–20 cents per gallon ethanol as
compared to only 2–4 cents per gallon ethanol for amylases used in the starch-to-
ethanol process, and ethanol production from sugarcane and molasses bypasses
this cost entirely. The cellulase requirement is also much higher than for amylases. The complex three-dimensional (crystalline) structure of cellulose slows
down the rate of hydrolysis dramatically; starch hydrolysis by amylases is 100
times faster than cellulose hydrolysis by cellulases under industrial processing
conditions.

The US Department of Energy (DOE), in 1999, signed contracts (worth
$17 million and $12.3 million, respectively) with Genencor International and
Novozymes Inc., to increase cellulase activity and bring down the cost of cellulose
enzymes. Novozymes has launched CellicTM, a new class of cellulases, which
is claimed to be more cost-effective and four times more active compared to the
previously produced cellulases (www.bioenergy.novozymes.com). On similar
lines, Genencor International has announced the launch of Accellerase 1,500 on
February 23, 2009. This is claimed to have significantly improved formulation and
higher activity resulting in higher ethanol yields and robust operation in a wider
variety of processes. The product is based on Genencor's patented gene expression
Technology.

Enzymatic hydrolysis is highly specific and can produce high yields of relatively pure glucose syrups, without generation of glucose degradation products.
Utility costs are low, as the hydrolysis occurs under mild reaction conditions. The
process is compatible with many pretreatment options, although purely physical
methods are typically not adequate (Graf and Koehler 2000; Sun and Cheng 2002).

Many experts see enzymatic hydrolysis as key to cost-effective ethanol production in the long run (US DOE 2003). Although acid processes are technically more mature, enzymatic processes have comparable projected costs and the potential of cost reductions as technology improves (Lynd et al. 1999).

The rate and extent of enzymatic hydrolysis of lignocellulosic biomass depends on enzyme loadings, hydrolysis periods, and structural features resulting from pretreatments. The extent of cellulose conversion is dictated by the degree of attachment of the cellulase enzymes to the three-dimensional cellulose surface. Crystallinity of cellulose affects the initial rate of enzymatic hydrolysis while presence of lignin and acetyl groups limits the extent of cellulose hydrolysis due to unproductive binding of cellulase with lignin and the acetyl groups (Ghose and Bisaria 1979). Thus, an efficient pretreatment strategy that decreases cellulose crystallinity and removes lignin to the maximum extent can significantly decrease hydrolysis time as well as cellulase loading. The problem of non-specific binding of cellulases to lignin can be circumvented by adding non-ionic surfactants like Tween 20. It has been shown that surfactant addition can increase the ethanol yield by 8 % and reduce cellulase loading by as much as 50 % while maintaining a constant ethanol yield. The proposed mechanism for this enhanced enzyme performance is the adsorption of the surfactant onto lignin surface, thereby preventing unproductive binding of the enzymes to lignin. The adsorption also helps in retaining the enzymes for recycle. Cheaper alternatives to non-ionic surfactants like Tween 20 can reduce the process cost considerably.

Another promising approach to enhancing enzymatic hydrolysis of pretreated biomass is the use of ultrasonic energy. The cavitational effect of ultrasound leads to biomass swelling and fragmentation, increasing the accessibility of cellulases to cellulosic substrate (Ebringerova and Hromadkova 2002). Intermittent ultrasonication of biomass before and after enzymatic hydrolysis has been shown to significantly increase the rate of reaction. Approximately 20 % increase in ethanol yield and 50 % decrease in enzyme loading were observed on intermittent exposure of simultaneous saccharification and fermentation processes of mixed waste office paper to ultrasonic energy under selected conditions (Banerjee et al. 2010). *Saccharomyces cerevisiae* is by far the most commonly used microbial species for industrial ethanol production from sugar- and starch-based raw materials and is well adapted to the industrial scenario. It produces ethanol with stoichiometric yields and tolerates a wide spectrum of inhibitors and elevated osmotic pressure. *Zymomonas mobilis* has been projected as the future ethanologen due to its high ethanol tolerance (up to 14 % v/v), energy efficiency, high ethanol yield (up to 97 % of theoretical), and high ethanol productivity. Complete substrate utilization is a prerequisite for economically competitive lignocellulosic fermentation processes. The native ethanologen *S. cerevisiae* is capable of fermenting only hexoses and cannot utilize pentoses like xylose, which is the main component of the hemicellulosic fraction of lignocellulose, and can contribute to as much as 30 % of the total biomass. *Z. mobilis* can only utilize glucose, fructose, and sucrose. Expanding the substrate range of whole-cell biocatalysts will greatly contribute to the economic feasibility of bioethanol production from renewable feedstock. The essential traits of a good lignocellulose-to-ethanol bioconverter are—utilization of both hexoses and pentoses; high ethanol yields and productivity; minimum by-product formation; high ethanol tolerance; and tolerance to other inhibitors formed during biomass pretreatment and hydrolysis. Four industrial benchmarks for ethanologenic strain development, which have the greatest influence on the price of lignocellulosic ethanol are—Process water economy, Inhibitor tolerance, Ethanol yield, Specific ethanol productivity (Banerjee et al. 2010).

The inability of *S. cerevisiae* and *Z. mobilis* to utilize pentose sugars necessitates the search for pentose utilizing strains. Two groups of microorganisms, i.e., enteric bacteria and some yeasts, are able to ferment pentoses, but with low ethanol yields (Bajpai and Margaritis 1982). Furthermore, xylose fermenting yeasts (*Pachysolen tannophilus*, *Candida shehatae*, and *Pichia stipitis*) are sensitive to high concentrations of ethanol (≥ 40 g/l), require microaerophilic conditions, are highly sensitive to inhibitors, and are not capable of fermenting xylose at low pH.

Due to the lack of a natural microorganism for efficient fermentation of lignocellulose-derived substrates, there has been emphasis on constructing an efficient organism through metabolic engineering of different organisms (Banerjee et al. 2010). Metabolic engineering, by virtue of the recent molecular biology tools, has generated recombinant organisms displaying attractive features for the bioconversion of lignocelluloses to ethanol. The three most promising microbial species that have been developed by metabolic engineering in the last two decades are *S. cerevisiae*, *Z. mobilis*, and *Escherichia coli*. The recombinant organisms developed for lignocellulose-to-ethanol process have been extensively reviewed in earlier works. Significant work on the metabolic engineering of *E. coli* has been completed. The incorporation and expression of pyruvate decarboxylase and alcohol dehydrogenase II genes from *Z. mobilis* into *E. coli* under the control of a common (*lac*) promoter results in high ethanol yield due to co-fermentation of hexose and pentose sugars. Xylose isomerase (XI), encoded by the *xylA* gene, catalyzes the isomerization of xylose to xylulose in bacteria and some fungi. Recent developments to improve XI activity and heterologous expression in *S. cerevisiae* through adaptation and metabolic manipulation have proven to be successful with ethanol yields approaching the theoretical maximum. Genes *xylA* and *xylB*, part of the xyl operon of *E. coli*, were introduced into *Pseudomonas putida* S12 under the transcriptional control of the constitutive *tac* promoter. This recombinant strain, after further laboratory evolution, could efficiently utilize the three, most abundant sugars in lignocellulose, glucose, xylose, and arabinose, as sole carbon sources. Whereas recombinant laboratory strains are useful for evaluating metabolic engineering strategies, they do not possess the robustness required in the industrial context. The strategies of producing industrial pentose fermenting strains involve the introduction of the initial xylose and arabinose utilization pathways and adaptation strategies, including random mutagenesis, evolutionary engineering, and breeding.

3.1.3.3 Fermentation

Fermentation of the hydrolyzed biomass can be carried out in a variety of different process configurations:

- Separate hydrolysis and fermentation (SHF)
- Simultaneous saccharification and fermentation (SSF)
- Simultaneous saccharification and co-fermentation (SSCF).

Enzymatic hydrolysis performed separately from fermentation step is known as separate hydrolysis and fermentation (SHF) (Wingren et al. 2003). The separation of hydrolysis and fermentation offers various processing advantages and opportunities. It enables enzymes to operate at higher temperature for increased performance and fermentation organisms to operate at moderate temperatures, optimizing the utilization of sugars. The most important process improvement made for the enzymatic hydrolysis of biomass is the introduction of simultaneous saccharification and fermentation (SSF), which has been improved to include the co-fermentation of multiple sugar substrates (Wingren et al. 2003). This approach combines the cellulase enzymes and fermenting microbes in one vessel. This enables a one-step process of sugar production and fermentation into ethanol. Simultaneous saccharification of both carbon polymer, cellulose to glucose; and hemicellulose to xylose and arabinose; and fermentation is carried out by recombinant yeast or the organism, which has the ability to utilize both C5 and C6 sugars. According to Alkasrawi et al. (2006), the mode of preparation of yeast must be carefully considered in SSF designing. A more robust strain will give substantial process advantages in terms of higher solid loading and possibility to recirculate the process stream, which results in increased energy demand and reduced fresh water utilization demand in process. Adaptation of yeast to the inhibitors present in the medium is an important factor for consideration in the design of SSF process. SSF combines enzymatic hydrolysis with ethanol fermentation to keep the concentration of glucose low. The accumulation of ethanol in the fermenter does not inhibit cellulase action as much as high concentration of glucose; so, SSF is good strategy for increasing the overall rate of cellulose-to-ethanol conversion (Tanaka 2006; Krishna et al. 2001; Kroumov et al. 2006). SSF gives higher ethanol yield while requiring lower amounts of enzyme because end-product inhibition from cellobiose and glucose formed during enzymatic hydrolysis is relieved by the yeast fermentation (Banat et al. 1998). However, it is not feasible for SSF to meet all the challenges at industrial level due to its low rate of cellulose hydrolysis and most microorganisms employed for ethanol fermentation cannot utilize all sugars derived after hydrolysis. To overcome this problem, the cellulolytic enzyme cocktail should be more stable in wide range of pH and temperature. Also, the fermenting microorganisms should be able to ferment a wide range of C5 and C6 sugars. Matthew et al. (2005) have found some promising ethanol producing bacteria, viz. recombinant *E. coli* K011, *Klebsiella oxytoca*, and *Zymomonas mobilis* for industrial exploitation. SSF process has now improved after including the co-fermentation of multiple sugar substrates present in the hydrolyzate. This new variant of SSF is known as simultaneous saccharification and co-fermentation (SSCF) (Wyman et al. 2005). SSF and SSCF are preferred over SHF, since both operations can be performed in the same tank resulting in lower cost, higher ethanol yield, and shorter processing time. The most upgraded form of biomass-to-ethanol conversion is consolidated bioprocessing (CBP)—featuring cellulose production, cellulose hydrolysis and fermentation in one step—is a highly integrated approach with outstanding potential (Lynd et al. 2005). It has the potential to provide the lowest cost route for biological conversion of cellulosic biomass to ethanol with high rate and desired yields.

Direct microbial conversion is a method of converting cellulosic biomass to ethanol in which both ethanol and all required enzymes are produced by a single microorganism. The potential advantage of direct microbial conversion is that a dedicated process step for the production of cellulase enzyme is not necessary. Cellulase enzyme production (or procurement) contributes significantly to the cost involved in enzymatic hydrolysis process. However, direct microbial conversion is not considered the leading process alternative. This is because there is no robust organism available that can produce cellulases or other cell wall degrading enzymes in conjunction with ethanol with a high yield. Singh and Kumar (1991) found that several strains of *Fusarium oxysporum* have the potential for converting not only D-xylose, but also cellulose to ethanol in a one-step process. Distinguishing features of *F. oxysporum* for ethanol production in comparison with other organisms are identified. These include the advantage of in situ cellulase production and cellulose fermentation, pentose fermentation, and the tolerance of sugars and ethanol. The main disadvantage of *F. oxysporum* is its slow conversion rate of sugars to ethanol as compared to yeast.

The comparison of the process conditions and performance of the different cellulose hydrolysis process shows that the dilute acid process has a low sugar yield (50–70 % of the theoretical maximum). The enzymatic hydrolysis has currently high yields (75–85 %), and improvements are still projected (85–95 %). Moreover, refraining from using acid may be better for the economics (cheaper construction materials, cutting operational costs), and the environment (no gypsum disposal).

The sugar syrup obtained after cellulosic hydrolysis is used for ethanol fermentation. A variety of microorganism—bacteria, yeast, or fungi—ferment sugars to ethanol under oxygen-free conditions. They do so to obtain energy and to grow. According to the reactions, the theoretical maximum yield is 0.51 kg ethanol and 0.49 kg carbon dioxide per kg sugar.

Methods for C6 sugar fermentation were already known. The ability to ferment pentoses along with hexoses is not widespread among microorganisms. *S. cereviseae* is capable of converting only hexose sugars to ethanol. The most promising yeasts that have the ability to use both C5 and C6 sugars are *Pichia stipitis*, *Candida shehatae*, and *Pachysolan tannophilus*. However, ethanol production from sugars derived from starch and sucrose has been commercially dominated by the yeast *S. cereviseae* (Tanaka 2006). Thermotolerant yeast could be more suitable for ethanol production at industrial level. In high temperature process, energy savings can be achieved through a reduction in cooling costs.

Considering this approach, Sree et al. (1999) developed solid state fermentation system for ethanol production from sweet sorghum and potato employing a thermotolerant *S. cereviseae* strain.

Currently, the research is focusing on developing recombinant yeast, which can greatly improve the ethanol production yield by metabolizing all form of sugars, and reduce the cost of operation. The efforts have been made by using two approaches. The first approach has been to genetically modify the yeast and other natural ethanologens additional pentose metabolic pathways. The second approach is to improve ethanol yields by genetic engineering in microorganisms that have the ability to ferment both hexoses and pentoses (Jeffries and Jin 2000; Dien et al. 2003). Jeffries and Jin (2000) compiled the developments toward the genetic engineering of yeast metabolism and concluded that strain selection through mutagenesis, adaptive evolution using quantitative metabolism models may help to further improve their ethanol production rates with increased productivities. Piskur et al. (2006) showed the developments in comparative genomics and bioinformatics to elucidate the high ethanol production mechanism from *Saccharomyces* sp. Though new technologies have greatly improved bioethanol production, yet there are still a lot of problems that have to be solved. The major problems include: maintaining a stable performance of genetically engineered yeast in commercial-scale fermentation operation (Ho et al. 1998, 1999), developing more efficient pretreatment technologies for lignocellulosic biomass, and integrating optimal component into economic ethanol production system (Dien et al. 2000). Fermentation can be performed as a batch, fed batch, or continuous process. The choice of most suitable process will depend upon the kinetic properties of microorganisms and type of lignocellulosic hydrolyzate in addition to process economics aspects. Traditionally, ethanol has been produced batch wise. At present, nearly, all of the fermentation ethanol industry uses the batch mode. In batch fermentation, the microorganism works in high substrate concentration initially and a high product concentration finally (Olsson and Han-Hagerdal 1996). The batch process is a multi-vessel process and allows flexible operation and easy control over the process. Generally, batch fermentation is characterized by low productivity with an intensive labor. For batch fermentation, elaborate preparatory procedures are needed; and because of the discontinuous start up and shut down operations, high labor costs are incurred. This inherent disadvantage and the low productivity offered by the batch process have led many commercial operators to consider the other fermentation methods like fed batch fermentation, continuous fermentation, immobilized cells, etc. In fed batch fermentation, the microorganism works at low substrate concentration with an increasing ethanol concentration during the course of fermentation process. Fed batch cultures often provide better yield and productivities than batch cultures for the production of microbial metabolites. For practical reasons, therefore, some continuous operations have been replaced by fed batch process. Keeping the low feed rate of substrate solution containing high concentration of fermentation inhibitors such as furfural, hydroxymethyl furfural, and phenolics, the inhibitory effect of these compounds to yeast can be reduced. Complete fermentation of an acid hydrolyzate of spruce, which was strongly inhibiting in batch fermentation, has been achieved without any detoxification treatment (Taherzadeh 1999). The productivity in fed batch fermentation is limited by the feed rate which, in turn, is limited by the cell mass concentration. The specific ethanol productivity has also been reported to decrease with increasing cell mass concentration (Palmqvist et al. 1996). Ideally, the cell density should be kept at a level providing maximum ethanol productivity and yield. Continuous fermentation can be performed in different kind of bioreactors—stirred tank reactors (single or series) or plug flow reactors. Continuous fermentation often gives a higher productivity than batch fermentation, but at low dilution rates which offers the highest productivities. Continuous operation offers ease of control and is less labor intensive than batch operation. However, contamination is more serious in this operation. Since the process must be interrupted, all the equipments must be cleaned, and the operation started again with the growth of new inoculum. The continuous process eliminates much of the unproductive time associated with cleaning, recharging, adjustment of media and sterilization. A high cell density of microbes in the continuous fermenter is locked in the exponential phase, which allows high productivity and overall short processing of 4–6 h as compared to the conventional batch fermentation (24–60 h). This results in substantial savings in labor and minimizes investment costs by achieving a given production level with a much smaller plant. A limitation to continuous fermentation is the difficulty of maintaining high cell concentration in the fermenter. The use of immobilized cells circumvents this difficulty. Immobilization by adhesion to a surface (electrostatic or covalent), entrapment in polymeric matrices, or retention by membranes has been successful for ethanol production from hexoses (Godia et al. 1987).

The applications of immobilized cells have made a significant progress in fuel ethanol production technology (Najafpour 1990; Abbi et al. 1996; Sree et al. 2000; Yamada et al. 2002). Immobilized cells offer rapid fermentation rates with high productivity—that is, large fermenter volumes of mash put through per day, without risk of cell washout. In continuous fermentation, the direct immobilization of intact cells helps to retain cells during transfer of broth into collecting vessel. Moreover, the loss of intracellular enzyme activity can be kept to a minimum level by avoiding the removal of cells from downstream products. Immobilization of microbial cells for fermentation has been developed to eliminate inhibition caused by high concentration of substrate and product and also to enhance ethanol productivity and yield.

3.1.3.4 Recovery of Ethanol

The product stream from fermentation is a mixture of ethanol, cell mass, and water. In this flow, ethanol from cellulosic biomass has likely lower product concentrations (≤ 5 wt %) than in ethanol from corn. The maximum concentration of ethanol tolerated by the microorganisms is about 10 wt % at 30 °C but decreases with increasing temperature. To maximize cellulase activity, the operation is rather at maximum temperature (37 °C), since the cost impact of cellulase production is high relative to distillation (Lynd 1996). On the processing side, slurries become difficult to handle when containing over 15 wt % solids, which also corresponds to 5 % ethanol (two-thirds carbohydrates, and <50 wt % conversion) (Lynd 1996). The first step is to recover the ethanol in a distillation or beer column, where most of the water remains with the solids part. The product (37 % ethanol) is then concentrated in a rectifying column to a concentration just below the azeotrope (95 %) (Wooley et al. 1999). Hydrated ethanol can be employed in E95 ICEVs (Wyman et al. 1993) or in FCVs (requires onboard reforming), but for mixtures with gasoline water-free (anhydrous) ethanol is required. One can further distill in the presence of an entrainer (e.g., benzene), dry by desiccants (e.g., corn grits), or use pervaporation or membranes (Lynd 1996). By recycling between distillation and dehydration, eventually 99.9 % of the ethanol in the beer is retained in the dry product (Wooley et al. 1999).

The main solid residual from the process is lignin. Its amount and quality differs with feedstock and the applied process. Production of co-products from lignin, such as high-octane hydrocarbon fuel additives, may be important to the competitiveness of the process (US DOE 2003). Lignin can replace phenol in the widely used phenol formaldehyde resins. Both production costs and market value of these products are complex. In corn-based ethanol plants, the stillage (20 % protein) is very valuable as animal feed.

3.1.3.5 Recirculation of Process Stream

The water consumption is reduced by recirculating process streams for use in the washing and hydrolysis steps (Palmqvist and Hahn-Hagerdal 2000). Recirculating part of the dilute ethanol stream from the fermenter can increase the ethanol concentration in the feed to the distillation stage. However, computer simulations have shown that recirculation of streams leads to the accumulation of nonvolatile inhibitory compounds (Galbe and Zacchi 1992; Palmqvist et al. 1996). To increase the ethanol productivity, cell recycling has been employed by several workers while retaining the simplicity of the batch process. Cell recycling generally does not increase the sugar consumption or ethanol production, but the time required for the fermentation can be reduced by 60–70 %. Schneider (1989) observed a reduction in ethanol production after third cell cycle and suggested the decrease in ethanol production was due to the limitations of oxygen and sugar as a result of an increase in cell density.

3.2 Production Costs

The production costs of bioethanol is variable (Table 3.2). It depends on the source of biomass. If the production costs for gasoline are 0.25 Euro/L, then this emphasizes the need to have governmental tax rebates in closing the price gap between

Table 3.2 Production costs of bioethanol

Biomass source	Production costs €/liter
US corn	0.42
Corn stover	0.45–0.58
EU wheat	0.27–0.43
EU sugar beet	0.32–0.54
Brazil sugarcane	0.16–0.28
Molasses (China)	0.24
Sweet sorghum (China)	0.22
Corn fiber (USA)	0.41
Wheat straw (USA)	0.44
Spruce (softwood)	0.44–0.63
Salix (hardwood)	0.48–0.71
Lignocellulose (biowaste)	0.11–0.32
Gasoline	0.25

Sassner et al. (2008) and Gnansounou (2008)

biofuel and fossil fuels. Economic drivers for the production and consumption of all biofuels are linked to the global price of oil. This is obviously a dynamic situation (with increasing oil prices improving the case for biofuels). It is apparent that for first-generation bioethanol feedstocks, Brazilian sugarcane represents one of the cheapest. Goldemberg (2007) estimated the production costs of bioethanol from sugar beet at 26.1 €/GJ, from sugarcane (Brazil) at 7.3 €/GJ; from sugarcane (USA and UK) at 12.3 €/GJ, from maize 9.3 €/GJ and from cellulose 20.3 €/GJ. EUBIA (2004) estimated 17.9 €/GJ for bioethanol from sugar beet, 16.7 €/GJ from wheat, 5.2 €/GJ from sugarcane (Brazil) and 13.6 €/GJ from maize. Ecofys (2003) estimated the production costs of bioethanol from straw at 26.0 €/GJ, from sugar beet at 27.9 €/GJ and from wheat at 29.8 €/GJ. Another study (AEA 2003) found 29.3 €/GJ for bioethanol from EU straw or beet pulp, 24.2 €/GJ from EU sugar beet, 21.3 €/GJ from EU wheat, 11.2 €/GJ from US maize, 9.0 €/GJ from Brazilian sugarcane in Brazil, and 31.1 €/GJ from Brazilian sugarcane in the UK. For bioethanol to be economically competitive with fossil fuels, production costs should be no greater than ~0.2 €/liter compared with gasoline.

The figures presented in table are approximations due to fluctuating raw material costs. For example, the US corn ethanol production costs are based on $4 bushel corn (32 lbs of starch and 2.8 gals of ethanol). The 2010 cost of sugarcane is at a historical high and current ethanol production costs from this feedstock are estimated at around $0.35 per liter. Lignocellulosic biomass costs are highly feedstock dependent (waste wood and paper costs will vary widely depending on locality and transport costs). Lignocellulose-to-ethanol production costs would be expected to become lower in the future as new technology improves the overall conversion processes. Biomass feedstock costs represent the predominant expenditure in bioethanol production, with first-generation feedstocks generally 50–80 % of total costs, while for lignocelluloses bioethanol processes, the feedstock costs are only ~40 % of total costs (Petrou and Pappis 2009).

3.3 Important Developments in the Production of Cellulosic Ethanol

Ethanol-from-cellulose (EFC) holds great potential due to the widespread availability, abundance, and relatively low cost of cellulosic materials. Significant investment into research, pilot, and demonstration plants is ongoing to develop commercially viable processes utilizing the biochemical and thermochemical conversion technologies for ethanol. Johnson et al. (2010) have reviewed the current status of commercial lignocellulosic ethanol production. Table 3.3 summarizes all known facilities as of February 2009, and these are shown geographically on Fig. 3.3 (Johnson et al. 2010).

Currently, the USA has a target of 136,260 million liters per year (ML/yr) of renewable fuels production by 2022. This target is only achievable with a majority of this renewable fuel coming from lignocellulosic material, such as corn stover, wood, switch grass, wheat straw, and purpose grown energy crops. Demonstration-scale cellulosic ethanol plants are under construction as part of the government's goal to make cellulosic ethanol cost competitive. The plants cover a wide variety of feedstocks, conversion technologies, and plant configurations to help identify viable technologies and processes for full-scale commercialization. All demonstration plants, which are sized at 10 % of a commercial-scale biorefinery, are expected to be operational soon. Commercial-scale plants are in the planning stages. Demonstration and commercial plants include—Abengoa—Alico, Alltech, American Energy Enterprises (AEE), Bluefire Ethanol, Coskata, Flambeau River Papers, Park Falls, Wisconsin, Fulcrum-Bioenergy, Sierra Biofuels Plant, ICM, Mascoma, The Wisconsin Rapids, Red Shield Environmental (RSE), Pacific Ethanol, The BioGasol process, Poet, Pure Energy & Raven BioFuels, Range Fuels, Verenium, Virent. Several efforts are underway in North America to commercially produce ethanol from wood and other cellulosic materials as a primary product. Table 3.4 partially summarizes these companies and their activities, which are in various states of progress.

Canadian government has set it sights on a target of 5 % renewable fuel in gasoline by 2010 and 2 % renewable fuel in diesel by 2012. To support this, the federal government has established funding of C$550 million dollars for pilot plants and process development and precommercial development, with a further C$500 million for demonstration-scale facilities and to assist with bridging the gap

Table 3.3 Lignocellulosic facilities as at February 2009

	Pilot/Demonstration	Commercial
Biochemical	25	9
Thermochemical	5	3

Pilot scale is R&D
Demonstration scale is < 10 ML/yr
Commercial scale is > 10 ML/yr
Johnson et al. (2010)

Table 3.4 Top 10 ethanol producers by capacity, March 2008 (existing production capacity—million gallons per year)

POET	1,208
Archer Daniels Midland (ADM)	1,070
VeraSun Energy Corporation	560
U.S. BioEnergy Corp.	420
Hawkeye Renewables	225
Aventine Renewable Energy	207
Abengoa Bioenergy Corp.	198
White Energy	148
Renew Energy	130
Cargill	120
All others	4,024
Total	8,310

Based on Yacobucc (2008)

between development and commercialization of cellulosic ethanol technologies. Facilities under construction or planned include: Iogen, lignol.

Details of facilities under construction or planned in the rest of the world are discussed below:

In South America, Brazil is today producing ~40 % of the world's ethanol from sugarcane. Yields vary from 6,600–7,500 L/ha which means production costs half those of US corn-based ethanol processors. Therefore, there has been relatively low interest in second-generation bioethanol. However, recent studies made by Brazilian National Development Bank (BNDES) together with FAO show that when fermenting bagasse and sugarcane the ethanol yield could reach 13,000 L/ha. The plants are located in Dedini, Sao Paulo, Brazil and in Chile.

In Europe, Abengoa, Spain—Abengoa Bioenergy has been operating a biomass-to-ethanol pilot plant since the end of 2007 at the Biocarburantes Castilla y León grain-ethanol plant in Babilafuente, near Salamanca in Spain; 5 ML/yr are produced from wheat and barley straw using enzymatic hydrolysis (glucose). A steam explosion pretreatment stage, from SunOpta, is currently being installed and will start up early 2009. In the second phase, it is intended to separate the lignin and pentose sugars as co-products. Chemrec, Pitea, Sweden—Chemrec has developed a gasification process to convert pulp mill black liquor (BL) to liquid fuels including ethanol. Other plants in Europe are: SEKAB, Ornskoldsvik, Sweden, Stora Enso, Varkaus, Finland UPM Kymmene, Finland, Elsam, Denmark Choren, Freiburg, Germany-DONG Energy, Denmark, BioGasol Denmark, TMO Renewables, Guildford, UK. In Japan, Australasia, Asia, following plants are under construction or planned. The BioEthanol Japan plant in Osaka Prefecture; Honda-RITE, Japan; Marubeni, Saraburi, Thailand; China Resources Alcohol Corporation; China; Mission NewEnergy, India; Ethtec, Australia—Ethtec, a Willmott Forests subsidiary; Pure Power, Singapore; LanzaTech, New Zealand; Scion, Rotorua, New Zealand.

A University of Florida researcher has developed a biotech "bug" that is capable of converting cellulosic biomass to ethanol. Lonnie Ingram, Director of the Florida Center for Renewable Chemicals and Fuels, has developed genetically engineered *E. coli* bacteria that can convert all types of sugar found in plant cell walls into fuel ethanol (Newswise 2005). The bacteria produce a high yield of ethanol from biomass such as sugarcane residues, rice hulls, forestry and wood wastes, and other organic materials. Ingram says he genetically engineered the *E. coli* organisms by cloning the unique genes needed to direct the digestion of sugars into ethanol, the same pathway found in yeast and higher plants. With the ethanol genes, he says that bacteria produce ethanol from biomass sugars with 90–95 % efficiency. Ingram began research in this area in 1985 and says that Dr. Nancy Ho of Purdue University has also made considerable progress in engineering yeasts to use in this process. "Reducing the cost and improving the efficiency of converting cellulosic materials into fermentable sugars is one of the keys to progress. The Department of Energy's National Renewable Energy Laboratory (NREL) has partnered with private biotech companies to make significant strides in this area." Ingram's University of Florida technology has become Landmark Patent No. 5,000,000 through the U.S. Department of Commerce. It is being commercialized with assistance from the Department of Energy, and BC International Corp., based in Dedham, Massachusetts, holds exclusive rights to use and license the engineered bacteria.

NREL and its partners say that the research conducted in this area is an important step toward realizing the potential of biorefineries (www.ethanol.org/documents/6-05_Cellulosic_Ethanol.pdf). Biorefineries, analogous to today's oil refineries, will use plant and waste materials to produce an array of fuels and chemicals—not just ethanol. Biorefineries will extend the value-added chain beyond the production of renewable fuel only. Progress toward a commercially viable biorefinery depends on the development of real-world processes for biomass conversion. With these new technologies for the production of cellulosic ethanol, its promise becomes closer to reality with each passing day.

US President in his State of the Union address on January 23, 2007, announced a proposed mandate for 35 billion gallons of ethanol by 2017. It is widely recognized that the maximum production of ethanol from corn starch is 15 billion gallons per year, implying a mandated production of some 20 billion gallons per year of cellulosic ethanol by 2017. The plan includes $2 billion dollars funding for cellulosic ethanol plants, with an additional $1.6 billion announced by the USDA on January 27, 2007.

On February 28, 2007, The Department of Energy (DOE) announced that six companies—Abengoa Bioenergy Biomass of Kansas, LLC ALICO, Inc. BlueFire Ethanol, Inc. Broin Companies, Iogen Biorefinery Partners, LLC, Range Fuels—will be awarded cellulosic ethanol grants to help with the construction of cellulosic ethanol biorefineries (RFA 2007b). The release of these grants is a major step forward for the cellulosic ethanol industry as many ethanol producers were awaiting this announcement to move forward with their plans. Renewable Fuels Association President Bob Dinneen said that "Cellulosic ethanol production is a must if we truly aspire to move this country toward a more diverse energy future. While corn will remain a key component of our ethanol industry, the kind of production necessary to greatly reduce gasoline consumption in this country can only be realized from the addition of cellulosic material as a feedstock. These grants are critical to bringing cellulosic ethanol to the commercial market and underscore the important partnership the federal government must have with the U.S. ethanol industry to achieve both our short-term and long-term energy goals." The grants are designed to help ethanol producers with the upfront capital costs associated with construction of cellulosic ethanol biorefineries. The stated DOE goal is to prove the feasibility of cellulosic ethanol technology. Recipients are eligible for up to $100 million and must show a 60 % industry/40 % government cost share. In addition, DOE is working on a loan guarantee program for cellulosic ethanol biorefineries as authorized in the 2005 energy bill. This program has seen significantly slower progress but is equally as important to helping companies construct cellulosic ethanol facilities often costing 4–5 times that of a traditional corn ethanol biorefinery.

Cellulosic ethanol is on track be cost competitive with corn-based ethanol by 2016, a development that could drive the fuel's production, according to an industry survey conducted by Bloomberg New Energy Finance. The survey focused on 11 major players in the cellulosic ethanol industry, all of which use a technique known as enzymatic hydrolysis to break down and convert the complex sugars in non-food crop matter, and a fermentation stage to turn the material into ethanol, BNEF said. Cellulosic ethanol cost 94 cents a liter to produce in 2012, about 40 % more than ethanol made from corn, BNEF said. That price gap will close by 2016, surveyed cellulosic ethanol producers predicted. Project capital expenditures, feedstock, and enzymes used in the production process are still the largest costs of running a cellulosic ethanol plant, the respondents said in the survey. But technology has pushed operating costs lower. For example, enzyme costs for a liter of cellulosic ethanol dropped 72 % between 2008 and 2012 due to technological improvements, BNEF said. Cellulosic ethanol producers will shift their focus from technology enhancements to logistical planning over the next 5–10 years in an effort to rein in capital costs, suggesting the industry is maturing, said BNEF's lead biofuel analyst Harry Boyle. Globally, there are 14 enzymatic hydrolysis pilots, nine demonstration-stage projects, and 10 semi-commercial-scale plants either announced, commissioned or due online shortly, according to the survey. Five of the semi-commercial plants are in the USA and more are expected to open in Brazil in the near future, BNEF said. A semi-commercial facility with a capacity of 90 million liters per year requires an initial capital outlay of about $290 million. By 2016, when second- and third-generation plants with capacities between 90 m and 125 m liters will be commissioned, initial capital costs per installed liter are expected to fall from $3 to $2 due to economies of scale and a reduction in over-engineering, BNEF said.

3.4 Energy Balances

Bioethanol produced from lignocellulosic biomass and other biowaste materials generally result in very favorable (i.e., positive) NER (net energy ratio) values. The definition of net energy value (NEV) is the difference between the energy in the fuel product (output energy) and the energy needed to produce the product (input energy). Ethanol has a net positive energy balance. It takes less than 35,000 BTUs of energy to turn corn into ethanol, while the ethanol offers at least 77,000 BTUs of energy which shows that ethanol's energy balance is clearly positive

Feedstock	Energy balance
Maize	1–2
Sugarcane	6.5–9.5
Sugar beet	1.1–2.3
Sweet sorghum	0.9–1.1
Lignocellulose	Highly dependent on feedstock, but generally highly positive
Gasoline	6

Table 3.5 Energy balances for bioethanol production from different feedstocks

Based on Walker (2010)

(Shapouri et al. 1995, 2002, 2003; Shapouri and McAloon 2004; Lorenz and Morris 1995; Wang et al. 1999; Kim and Dale 2004a, b; Farrell et al. 2006) and an extremely high petroleum/fossil energy displacement ratio.

A similar useful parameter in this regard is the Net Energy Balance (NEB), which is the ratio of the ethanol energy produced to the total energy consumed (in biomass growth, processing and biofuel production). Table 3.5 presents energy balances from the production of bioethanol from sugarcane, maize, and lignocellulose, and it is apparent that of the first-generation biomass sources, sugarcane represents the most favorable feedstock with respect to energy balance. Energy balance values <1 mean that bioethanol production is unfeasible from energetic standpoint and is indicative of excess of fossil energy used to produce bioethanol. For maize (corn) ethanol processes in North America (USA), typical values are 1–2:1, while for sugarcane ethanol processes in South America (Brazil), typical values are 5–10:1. Figures are variable due to different geographic, climatic, and agricultural reasons, but for Brazilian sugarcane ethanol operations, a typical energy balance of 8 (i.e., 8 times energy production in comparison with inputs) and GHG reductions of 90 % (compared to only 30 % for ethanol from corn) are achievable (Amorim et al. 2009; Basso and Rosa 2010). Brazilian bioethanol plants that combust residual bagasse to steam for electricity generation have very favorable energy balances. Brazil is thus considered to be a sustainable biofuel producer. Calculations of energy balances in bioethanol production depend on several factors, for example, whether or not fossil fuel usage in agronomic practices and co-generation of energy from by-products are included. Nevertheless, there is scope to reduce energy inputs from the bioprocessing (rather than biomass cultivation) perspective, particularly through adoption of modern biotechnology (Mousdale 2008)

The advances in the farming community as well as technological advances in the production of ethanol have led to positive returns in the energy balance of ethanol. Studies have shown that the ethanol energy balance is improving by the year (Wang 2005, 1999; Shapouri et al. 1995, 2002, 2003; Lorenz and Morris 1995; Wang et al. 1999; Morris 1995). The energy output to energy input ratio for converting irrigated corn to ethanol is now 1.67–1. According to USDA report, 1995, the corn ethanol energy balance had a gain of 24 %. That same report was revisited the next year, the authors concluded the ratio had risen to 34 %. This number is reinforced by a 2002 report. The report concluded that ethanol production is energy efficient because it yields 34 % more energy than is used. The USDA in June 2004, again looked at this issue and determined that ethanol continues to be more efficient and now provides the aforementioned 1.67–1 gain in energy. Many advances have led to the surge in ethanol production efficiency. One key issue is the ability to produce more gallons of ethanol per bushel of corn. In the early 1990s, plants were able to produce about 2.5 gallons of ethanol per bushel. That number has since increased to between 2.7 and 2.8 gallons per bushel today.

Crop with a higher sugar content than corn, such as sugar beets, would result in production with a much higher positive net energy balance (www1.eere.energy. gov/biomass/net_energy_balance.html). If corn farmers use state of the art, energy efficient farming techniques, and ethanol plants use state of the art production processes, then the amount of energy contained in a gallon of ethanol and the other co-products is more than twice the energy used to grow the corn and convert it into ethanol. These studies indicated an industry average net energy gain of 1.38–1. The industry-best existing production net energy ratio was 2.09–1. If farmers and industry were to use all the best technologies and practices, the net energy ratio would be 2.51–1. In other words, the production of ethanol would result in more than 2–1/2 times the available energy than it took to produce it. A 1999 study by Argonne National Laboratory found the energy balance of cellulosic ethanol to be in excess of 60,000 Btu per gallon (Wang 1999). Given that feedstocks for cellulosic ethanol are essentially waste products like corn stover, rice bagasse, forest thinnings, or even municipal waste, there are relatively few chemical and energy inputs that go into the farming of feedstocks for cellulosic ethanol. A secondary factor, although to a much lesser extent, is the fact that cellulosic ethanol plants will presumably produce "extra" energy that can be fed into the power grid. Doing so will effectively displace the use of electricity produced in power plants, which for the most part rely upon fossil fuels. Ethanol's net energy value published by different researchers (Shapouri et al. 1995, 2002, 2003; Lorenz and Morris 1995; Wang et al. 1999; Pimental 2002) is shown in Table 3.6.

Table 3.6 Ethanol's net energy value reported by different researchers

NEV (Btu)
+20,436 (HHV)
+30,589 (HHV)
+29,826 (LHV)
+22,500 (LHV)
−33,562 (LHV)
+21,105 (HHV)
+23,866 to +35,463 (LHV)
+17,508
−22,300
+21,105
+30,258 (LHV)

Based on Shapouri et al. (1995, 2002, 2003, Shapouri and McAloon 2004), Lorenz and Morris (1995), Wang et al. (1999)

References

Abbas C (2010) Going against the grain: food versus fuel uses of cereals. In: Walker GM, Hughes PS (eds) Distilled spirits. New horizons: energy, environment and enlightenment. Proceedings of the worldwide distilled spirits conference, Edinburgh, 2008. Nottingham University Press

Abbi M, Kuhad RC, Singh A (1996) Bioconversion of pentose sugars to ethanol by free and immobilized cells of *Candida shehatae* NCL-3501: fermentation behaviour. Proc Biochem 31:55–560

AEA (2003) International resource costs of biodiesel and bioethanol. AEA Technology in commission of the UK Department of Transport. Available at http://www.dft.gov.uk/pgr/roads/environment/research/cqvcf/internationalresourcecostsof3833?page=3

Alkasrawi A, Rudolf A, Lid'en G, Zacchi G (2006) Influence of strain and cultivation procedure on the performance of simultaneous saccharification and fermentation of steam pretreated spruce. Enzyme Microb Technol 38:279–286

Amorim HV, Basso LC, Lopes ML (2009) Sugar can juice and molasses, beet molasses and sweet sorghum: composition and usage. In: The alcohol textbook, 5th edn. Nottingham University Press, pp 39–46

Badger PC (2002) Ethanol from cellulose: a general review. In: Janick J, Whipkey A (eds) Trends in new crops and new uses. ASHS Press, Alexandria, VA, USA

Bajpai P, Margaritis A (1982) Direct fermentation of D-xylose to ethanol by Kluyveromyces marxianus strains. Appl Environ Microbiol 44(5):1039–1041

Bajpai P (2007) Bioethanol. PIRA Technology Report, Smithers PIRA, UK

Banat IM, Nigam P, Singh D, Marchant P, McHale AP (1998) Ethanol production at elevated temperatures and alcohol concentrations. Part I: yeasts in general. World J Microbiol Biotechnol 14:809–821

Banerjee S, Mudliar S, Sen R et al (2010) Commercializing lignocellulosic bioethanol: technology bottlenecks and possible remedies. Biofuels, Bioprod Biorefin 4:77–93

Basso LC, Rosa CA (2010) Sugar cane for potable and fuel ethanol. In: Proceedings of the worldwide distilled spirits conference, Edinburgh, 2008. Nottingham University Press (in press)

Cadoche L, Lopez GD (1989) Assessment of size reduction as a preliminary step in the production of ethanol from lignocellulosic wastes. Biol Wastes 30:153–157

Chandel AK, Chan ES, Rudravaram R, Lakshmi M, Venkateswar R, Ravindra P (2007) Economics and environmental impact of bioethanol production technologies: an appraisal. Biotechnol Mole Boil Rev 2(1):014–032

Damaso MCT, Castro Mde, Castro RM, Andrade MC, Pereira N (2004) Application of xylanase from *Thermomyces lanuginosus* IOC-4145 for enzymatic hydrolysis of corn cob and sugarcane bagasse. Appl Biochem Biotechnol 115:1003–1012

Department of Energy (2003) Advanced bioethanol technology, website: www.ott.doe.gov/biofuels/, US Department of Energy, Office of Energy Efficiency and Renewable Energy, Office of Transportation Technologies, Washington DC, USA

Dien BS, Nichols NN, O'Bryan PJ, Bothast RJ (2000) Development of new ethanologenic *Escherichia coli* strains for fermentation of lignocellulosic biomass. Appl Biochem Biotechnol 84(86):181–196

Dien BS, Cotta MA, Jeffries TW (2003) Bacteria engineered for fuel ethanol production current status. Appl Microbiol Biotechnol 63:258–266

Dien BS, Jung HJG, Vogel KP, Casler MD, Lamb JAFS, Iten L, Mitchell RB, Sarath G (2006). Chemical composition and response to dilute acid pretreatment and enzymatic saccharification of alfalfa, reed canary grass and switch grass. Biomass Bioenergy 30(10):880–891

Ebringerova A, Hromadkova Z (2002) Effect of ultrasound on the extractability of corn bran hemicelluloses. Ultrason Sonochem 9(4):225–229

Ecofys (2003) Biofuels in the Dutch market: a fact-finding study, Utrecht, the Netherlands, report no. 2GAVE-03.12

Eggeman T, Elander TR (2005) Process and economic analysis of pretreatment technologies. Biores Technol 8:2019–2025

EUBIA (2004) Biofuel for transport, G. Grassi, European Biomass Industry Association, accessed at http://www.eubia.org/

Fan LT, Gharpuray MM, Lee YH (1987) Cellulose hydrolysis biotechnology monographs. Springer, Berlin, p 57

Farrell AE, Plevin RJ, Turner BT, Jones AD, Hare MO, Kammen DM (2006) Ethanol can contribute to energy and environmental goals. Science 311 (Jan 27)

Ferguson A (2003) Implication of the USDA 2002 update on ethanol from corn, vol 3, no 1. The Optimum Population Trust, Manchester, UK

Galbe M, Zacchi G (1992) Simulation of ethanol production processes based on enzymatic hydrolysis of lignocellulosic materials using Aspen. Appl Biochem Biotechnol 34–35:93–104

Ghose TK, Bisaria VS (1979) Studies on mechanism of enzymatic hydrolysis of cellulosic substances. Biotechnol Bioeng 21:131–146

Gnansounou E (2008) Fuel ethanol. Current status and outlook. In: Pandey A (ed) Handbook of plant-based biofuels. CRC Press, Boca Raton, pp 57–71

Godia F, Casas C, Sola C (1987) A survey of continuous ethanol fermentation systems using immobilized cells. Process Biochem 22–22:43–48

Goldemberg J (2007) Ethanol for a sustainable energy future. Science 315:808–810

Graf A, Koehler T (2000) Oregon cellulose-ethanol study. Oregon Office of Energy, Salem, OR USA, p 30 + appendices

Gray KA, Zhao L, Emptage M (2006) Bioethanol. Curr Opin Chem Biol 10:141–146

Gregg DJ, Saddler JN (1996) Factors affecting cellulose hydrolysis and the potential of enzyme recycle to enhance the efficiency of an integrated wood to ethanol process. Biotechnol Bioeng 51:375–383

Ho NWY, Chen Z, Brainard AP (1998) Genetically engineered *Saccharomyces* yeast capable of effective co-fermentation of glucose and xylose. Appl Environ Microbiol 64:1852–1859

Ho NWY, Chen Z, Brainard A, Sedlak M (1999) Successful design and development of genetically engineered *Saccharomyces* yeasts for effective co fermentation of glucose and xylose from cellulosic biomass to fuel ethanol. Adv Biochem Eng Biotechnol 65:164–192

Hsu TA, Ladisch MR, Tsao GT (1980) Alcohol from cellulose. Chem Technol 10(5):315–319

Itoh H, Wada M, Honda Y, Kuwahara M, Watanabe T (2003) Bioorganosolve pretreatments for simultaneous saccharification and fermentation of beech wood by ethanolysis and white-rot fungi. J Biotechnol 103(3):273–280

Jeffries TW, Jin YS (2000) Ethanol and thermotolerance in the bioconversion of xylose by yeasts. Adv Appl Microbiol 47:221–268

Johnson T, Johnson B, Scott-Kerr C, Kiviaho J (2010) Bioethanol—status report on bioethanol production from wood and other lignocellulosic feedstocks. In: 63rd Appita annual conference and exhibition, Melbourne, 19–22 Apr 2009

Keller FA, Hamilton JE, Nguyen QA (2003) Microbial pretreatment of biomass: potential for reducing severity of thermochemical biomass pretreatment. Appl Biochem Biotechnol 27–41:105–108

Kim S, Dale BE (2004a) Cumulative energy and global warming impact from the production of biomass for biobased products. J Indust Ecol 7(3–4):147–162

Kim S, Dale EB (2004b) Global potential bioethanol production from wasted crops and crop residues. Biomass Bioenergy 26:361–375

Kim S, Holtzapple MT (2006) Lime pretreatment and enzymatic hydrolysis of corn stover. Bioresour Technol 96:1994–2006

Kim HT, Kim JS, Sunwoo C, Lee YY (2003) Pretreatment of corn stover by aqueous ammonia. Biores Technol 90:39–47

Klinke HB, Thomsen AB, Ahring BK (2004) Inhibition of ethanol producing yeast and bacteria by degradation products produced during pretreatment of biomass. Appl Microbiol Biotechnol 66:10–26

Krishna SH, Reddy TJ, Chowdary GV (2001) Simultaneous saccharification and fermentation of lignocellulosic wastes to ethanol using thermotolerant yeast. Bioresour Technol 77:193–196

Kroumov AD, M'odenes AN, de Araujo Tait MC (2006) Development of new unstructured model for simultaneous saccharification and fermentation of starch to ethanol by recombinant strain. Biochem Eng J 28:243–255

Kuhad RC, Singh A, Ericksson KE (1997) Microorganisms and enzymes involved in the degradation of plant fiber cell walls. Adv Biochem Eng Biotechnol 57:45–125

Ladisch MR (2003) Apollo program for biomass liquids what will it take? Universiteit Utrecht Copernicus Institute, Science Technology Society. Available at: www.purdue.edu/energysummit/presentations/ladisch_purdue.pdf

Larsson S, Palmqvist E, Hahn Hagerdal B, Tengborg C, Stenberg K, Zacchi G, Nilvevrant NO (1999) The generation of fermentation inhibitors during dilute acid hydrolysis of soft wood. Enzyme MicrobTechnol 24:151–159

Lopez MJ, Nichols NN, Dien BS, Moreno J, Bothast RJ (2004) Isolation of microorganisms for biological detoxification of lignocellulosic hydrolysates. Appl Microb Biotechnol 64:125–131

Lorenz D, Morris D (1995) How much energy does it take to make a gallon of ethanol?, Institute for Local Self-Reliance (Aug 1995)

Lynd LR (1996) Overview and evaluation of fuel ethanol from cellulosic biomass: technology, economics, the environment, and policy. Annu Rev Energy Environ 21:403–465

Lynd LR, Wyman CE, Gerngross TU (1999) Biocommodity engineering. Dartmouth College/Thayer School of Engineering, Hanover

Lynd LR, van Zyl WH, McBride JE, Laser M (2005) Consolidated bioprocessing of cellulosic biomass: an update. Curr Opin Biotechnol 16:577–583

Martín C, Galbe M, Wahlbom CF, Hägerdal BH, Jönsson LJ (2002) Ethanol production from enzymatic hydrolysates of sugarcane bagasse using recombinant xylose-utilising *Saccharomyces cerevisiae*. Enz Microb Technol 31:274–282

Martinez A, Rodriguez ME, York SW, Preston JF, Ingram LO (2000) Effects of Ca(OH)2 treatments ("overliming") on the composition and toxicity of bagasse hemicellulose hydrolysates. Biotechnol Bioeng 69:526–536

Matthew H, Ashley O, Brian K, Alisa E, Benjamin JS (2005) Wine making 101

Morris D (1995) How much energy does it take to make a gallon of ethanol? Institute for Local Self-Reliance. Available at: www.carbohydrateeconomy.org/library/admin/uploadedfiles/How_Much_Energy_Does_it_Take_to_Make_a_Gallon_.html-33k

Mosier N, Wyman C, Dale B, Elander R, Lee YY, Holtzapple M, Ladisch M (2005) Features of promising technologies for pretreatment of lignocellulosic biomass. Bioresour Technol 96:673–686

Mousdale DM (2008) Biofuels. Biotechnology, chemistry and sustainable development. CRC Press, Boca Raton

Nair SU, Ramachandran S, Pandey A (2008) Bioethanol from starch biomass. Part 1. Production of starch saccharifying enzymes. In: Pandey A (ed) Handbook of plant-based biofuels. CRC Press, Boca Raton, pp 87–103

Najafpour GD (1990) Immobilization of microbial cells for the production of organic acids. J Sci Islam Repub Iran 1:172–176

Newswise (2005) Biomass-to-ethanol technology could help replace half of U.S. auto fuel. http://newswise.com/articles/view/511547

Nigam JN (2002) Bioconversion of water-hyacinth (*Eichhornia crassipes*) hemicellulose acid hydrolysate to motor fuel ethanol by xylose–fermenting yeast. J Biotechnol 97:107–116

Nilvebrant N, Reimann A, Larsson S, Jonsson LJ (2001) Detoxification of lignocellulose hydrolysates with ion exchange resins. Appl Biochem Biotechnol 91–93:35–49

O'Brien D, Woolverton M (2009) Recent trends in U.S. wet and dry corn milling production. AgMRC Renewable Newsletter. http://www.agmrc.org/renewable_energy/

Olsson L, Hahn-Hägerdal B (1996) Fermentation of lignocellulosic hydrolysates for ethanol production. Enzyme Microb Technol 18:312–331

Olsson L, Jorgensen H, Krogh KBR, Roca C (2005) Bioethanol production from lignocellulosic material. In: Dumitriu S (ed) Polysaccharides. Structural diversity and functional versatility. 2nd edn, Chap. 42. Marcel Dekker, New York, pp 957–993

Palmqvist E, Hahn-Hagerdal B (2000) Fermentation of lignocellulosic hydrolysates. I: inhibition and detoxification and II: inhibitors and mechanisms of inhibition. Bioresour Technol 74:17–33

Palmqvist E, Hahn-Hägerdal B, Galbe M, Zacchi G (1996) The effect of water-soluble inhibitors from steam-pretreated willow on enzymatic hydrolysis and ethanol fermentation. Enzyme Microb Technol 19:470–476

Petrou EC, Pappis CP (2009) Biofuels: a survey on pros and cons. Energy Fuels 23:1055–1066

Pimentel D (2002) Limits of biomass utilization. Encyclopedia of physical science and technology. 3rd edn, vol 2. Academic Press. pp 159–171

Piskur J, Rozpedowska E, Polakova S, Merico A, Compagno C (2006) How did *Saccharomyces* evolve to become a good brewer? Trends Genet 22(4):183–186

RFA—Renewable Fuels Association (2007a) Ethanol industry outlook: building new horizons. Available at: http://www.ethanolrfa.org/objects/pdf/outlook/RFA_Outlook_2007.pdf

RFA—Renewable Fuels Association (2007b) Cellulosic ethanol grants provide much needed boost to fledgling technology. Available at: www.biofuelsjournal.com/articles/RFA

RFA—Renewable Fuels Association (2013) Battling for the barrell ethanolrfa.3cdn.net/dc207800043a5aa5aa_y5im6rokb.pdf

Saha BC, Iten LB, Cotta MA, Wu YV (2005). Dilute acid pretreatment, enzymatic saccharification and fermentation of wheat straw to ethanol. Proc Biochem 40:3693–3700

Sassner P, Galbe M, Zacchi G (2008) Techno-economic evaluation of bioethanol production from three different lignocellulosic materials. Biomass Bioenergy 32:422–430

Schneider HI (1989) Conversion of pentoses to ethanol by yeast and fungi. Crit Rev Biotechnol 9:1–40

Senthilkumar V, Gunasekaran P (2008) Bioethanol from biomass. Production of ethanol from molasses. In: Pandey A (ed) Handbook of plant-based biofuels. CRC Press, Boca Raton, pp 73–86

Shapouri H, McAloon A (2004) The 2001 net energy balance of corn ethanol. U.S Department of Agriculture, Washington, DC

Shapouri H, Duffield JA, Graboski MS (1995) Estimating the net energy balance of corn ethanol. AER-721. USDA Economic Research Service, Washington, DC

Shapouri H, Duffield JA, Wang M (2002) The energy balance of corn ethanol: an update. AER-814. USDA Office of the Chief Economist, Washington, DC

Shapouri H, Duffield JA, Wang M (2003) The energy balance of corn ethanol revisited. Am Soc Agric Eng 46(4):959–968

Sheehan J, Aden A, Paustian K, Killian K, Brenner J, Walsh M, Nelson R (2003) Energy and environmental aspects of using corn stover for fuel ethanol. J Ind Ecol 7:117–146

Singh A, Kumar PK (1991) *Fusarium oxysporum*: status in bioethanol production. Crit Rev Biotechnol 11(2):129–147

Sree NK, Sridhar M, Suresh K, Rao LV, Pandey A (1999) Ethanol production in solid substrate fermentation using thermotolerant yeast. Proc Biochem 34:115–119

Sree NK, Sridhar M, Suresh K, Banat IM, Rao LV (2000) Isolation of thermotolerant, osmotolerant, flocculating *Saccharomyces cerevisiae* for ethanol production. Biores Technol 72:43–46

Sun Y, Cheng J (2002) Hydrolysis of lignocellulosic materials for ethanol production: a review. Bioresour Technol 83:1–11

Taherjadeh M (1999) Ethanol from lignocellulose: physiological effects of inhibitors and fermentation strategies. Ph.D. Thesis. Lund University, Lund, Sweden

Tanaka L (2006) Ethanol fermentation from biomass resources: current state and prospects. Appl Microbiol Biotechnol 69:627–642

Tucker MP, Kim KH, Newman MM, Nguyen QA (2003) Effects of temperature and moisture on dilute-acid steam explosion pretreatment of corn stover and cellulase enzyme digestibility. Appl Biochem Biotechnol 10:105–108

von Sivers M, Zacchi G, Olsson L, Hahn-Hägerdal B (1994) Cost analysis of ethanol production from willow using recombinant *Escherichia coli*. Biotechnol Prog 10:555–560

Walker GM (2010) Bioethanol: science and technology of fuel alcohol. Ventus Publishing ApS, ISBN 978-87-7681-681-0

Wang M (1999) Argonne National Laboratory, biofuels: energy balance. Available at: http://www.iowacorn.org/ethanol/documents/energy_balance_000.pdf

Wang M (2005) An update of energy and greenhouse emission impacts of fuel ethanol. Center for Transportation Research Argonne National Laboratory, The 10th Annual National Ethanol Conference Scottsdale, AZ. Available at: http://www.ethanol-gec.org/netenergy/UpdateEnergyGreenhouse.pdf

Wang M, Saricks C, Santini D (1999) Effects of fuel ethanol on fuel-cycle energy and greenhouse gas emissions. Argonne National Laboratory. ANL/ESD-38, p 39. Available at: http://www.transportation.anl.gov/pdfs/TA/58.pdf

Wingren A, Galbe M, Zacchi G (2003) Techno-economic evaluation of producing ethanol from softwood: comparison of SSF and SHF and identification of bottlenecks. Biotechnol Prog 19:1109–1117

Wooley R, Ruth M, Sheehan J, Ibsen K, Majdeski H and Galvez A (1999) Lignocellulosic biomass to ethanol—process design and economics utilizing co-current dilute acid prehydrolysis and enzymatic hyrolysis—current and futuristic scenarios. National Renewable Energy Laboratory, Golden Colorade USA, p 72 + annexes

Wyman CE, Bain RL, Hinman ND, Stevens DJ (1993) Ethanol and methanol from cellulosic biomass. In: Johansson TB, Kelly H, Reddy AKN, Williams RH, Burnham L (eds) Renewable energy, sources for fuels and electricity. Island Press, Washington DC, pp 865–923

Wyman CE, Dale BE, Elander RT, Holtzapple M, Ladisch MR, Lee YY (2005) Comparative sugar recovery data from laboratory scale application of leading pretreatment technologies to corn stover. Biores Technol 96(18):2026–2032

Yacobucc (2008) CRS report for congress fuel ethanol: background and public policy issues. Updated 24 Apr 2008

Yacobucci B, Womach J (2003) Fuel ethanol: background and public policy issues. Library of Congress, Washington DC. Available at: http://www.ethanol-gec.org/information/briefing/1.pdf

Yamada T, Fatigati MA, Zhang M (2002) Performance of immobilized Zymomonas mobilis 31821 (pZB5) on actual hydrolysates produced by Arkenol technology. Appl Biochem Biotechnol 98:899–907

Zhang S, Marechal F, Gassner M, Perin-Levasseur Z, Qi W, Ren Z, Yan Y, Farvat D (2009) Process modelling and integration of fuel ethanol production from lignocellulosic biomass based on double acid hydrolysis. Energy Fuels. doi:10.1021/ef801027x

Chapter 4
Ethanol Markets

The ethanol market is driven by a number of factors including oil prices, national energy and foreign policy, federal and state tax incentives, and technological developments in the future. As expected, as oil prices increase, there is often a shift in the demand for ethanol as a biofuel. However, the market is often limited by the amount that can be blended with gasoline for conventional engines and the limited availability and use of flex-fuel vehicles. The limited availability of distribution points for ethanol-based fuel is also a consideration. The development of biofuel vehicles must go hand in hand with the development of the fuel distribution system. The ethanol market is also influenced by national and state energy policies with some states mandating ethanol content, sometimes without mandatory labeling of the ethanol content. State and federal tax incentives have a strong influence on the market and production of ethanol in the United States. Needless to say, many corn-to-ethanol plants would not be profitable without these incentives. It is expected that technological changes, especially the use of different biomass feedstocks such as wood and lignocellulosic materials, will have a significant effect on the market. Even with these considerations, the world's ethanol production is expected to increase significantly, with the United States and Brazil being the leading producers in the world, perhaps reaching as high as 28 billion gallons. Emerging market and production in Cuba, Asia, and Latin America will also contribute to the growth of the ethanol market. It is expected that much of this growth in ethanol production will need to be in the form of cellulosic ethanol so that the agricultural and food market (with respect to corn) is not significantly disrupted.

Currently, the largest market for ethanol is transportation fuel for passenger vehicles. Ethanol can be used in passenger vehicles, in any combination of blends, up to 10 % without any vehicle modifications. This includes E-10, commonly referred to as "gasohol," oxygenated fuel, and reformulated gasoline (RFG). Higher blends, such as E85, can only be used in modified vehicles. E85

Some excerpts taken from Bajpai (2007). PIRA Technology Report on Bioethanol with kind permission from Smithers PIRA

Table 4.1 Advantages and disadvantages of E85

Advantages	Disadvantages
Domestically produced, reducing use of imported petroleum	Can only be used in flex-fuel vehicles
Lower emissions of air pollutants	Lower energy content, resulting in fewer miles per gallon
More resistant to engine knock	Limited availability
Added vehicle cost is very small	Currently expensive to produce

Based on http://www.dingscompletecarcare.com/Go_Green.htm

flexible-fuel vehicles are manufactured by all three major automobile manufacturers and are modified to run on ethanol blends up to 85 %. Table 4.1 shows advantages and disadvantages of E85. There are also additional niche markets for ethanol-blended fuels (Launder 1999).

Ethanol, with its high-octane property, has established its role as an octane enhancer. Blending ethanol at 10 % raises the gasoline's octane level by an average of three octane points. As an octane enhancer, ethanol had to compete with other octane enhancers. In the early 1980s, methanol blends gained a short popularity but soon disappeared from the market due to public health risks associated with methanol's toxicity as well as to its undesirable corrosive effects on engines and pipelines. While methanol blends did not experience a wide market success, a methanol-derived ether called MTBE, also used to raise octane content, became popular as a blending agent. MTBE, first produced in 1979 in the United States and Europe, was developed primarily as an octane enhancer by combining isobutylene and methanol. Unlike ethanol, MTBE can easily be blended with gasoline at the refinery and transported by pipeline.

4.1 Transportation Fuel

Countries like USA and Brazil have long promoted domestic ethanol production. Ethanol/gasoline blends were promoted as an environmentally driven practice, in the USA initially as an octane enhancer to replace lead. Ethanol is also used as an oxygenate in clean-burning gasoline to reduce vehicle exhaust emissions. In USA, ethanol supplies today account for about 1 % of the highway motor vehicle fuel market, in the form of a gasoline blending component. At present, most of this ethanol is used in a 10 % blend with gasoline which is commonly referred to as "gasohol," a term which is being replaced with "ethanol/gasoline blends" or "E10." In some areas, lower percentage blends, containing 5.7 or 7.7 % ethanol, are also being used to match to air quality regulations affecting the oxygen content of reformulated gasoline (CFDC 1999). The 5.7 % blend is California's formulation used to meet a 2 % by weight federal oxygenate requirement in phase II gasoline. 5 % bioethanol blend does not require any engine modification and is covered by vehicle warranties. Combined with gasoline, ethanol increases octane levels while also promoting more complete fuel burning that reduces harmful exhaust emissions such as carbon monoxide and hydrocarbons (IEA 2004, 2006). In addition to ethanol/gasoline blend markets, ethanol has other motor fuel applications including

1. Use as E85, 85 % ethanol and 15 % gasoline
2. Use as E100, 100 % ethanol with or without a fuel additive
3. Use in oxydiesel, typically a blend of 80 % diesel fuel, 10 % ethanol and 10 % additives and blending agents

There are also smaller niche markets such as fuel cell applications, E diesel (a cleaner burning diesel fuel containing up to 15 % ethanol), aviation, where ethanol can be utilized (Launder 1999). Ethanol is used in the chemical industry for making variety of basic and intermediate chemicals. Bioethanol could be of great economic and environmental interest in developing countries, especially for cooking and lighting as a substitution for LPG (particularly for remote locations). Power and heat segment may be much larger market than the transport for bioethanol.

There are several common ethanol fuel mixtures in use around the world. The use of pure hydrous or anhydrous ethanol in ICE is only possible if the engine is designed or modified for that purpose. Anhydrous ethanol can be blended with petrol in various ratios for use in unmodified gasoline engines and with minor modifications can also be used with a higher content of ethanol. Ethanol fuel mixtures have "E" numbers which describe the percentage of ethanol in the mixture by volume, for example, E85 is 85 % anhydrous ethanol and 15 % gasoline. Low ethanol blends, from E5 to E25, are also known as *gasohol*, though internationally the most common use of the term gasohol refers to the E10 blend. Blends of E10 or less are used in more than twenty countries around the world, led by the United States, where ethanol represented 10 % of the US gasoline fuel supply in 2011. Blends from E20 to E25 have been used in Brazil since the late 1970s (ANFAVEA 2006). E85 is commonly used in the USA and Europe for flexible-fuel vehicles. Hydrous ethanol or E100 is used in Brazilian neat ethanol vehicles and flex-fuel light vehicles and in hydrous E15 called hE15 for modern petrol cars in the Netherlands. Table 4.2 shows ethanol blends used in Brazil.

The Energy Policy Act of 2005 established a Renewable Fuels Standard (RFS). This standard requires the use of 4.0 billion gallons of renewable fuels in 2006, increasing each year to 7.5 billion gallons in 2012. Most of this requirement is expected to be met with ethanol. In USA, approximately 3.4 billion gallons of ethanol were consumed in 2004.

Ethanol can be used as a replacement for MTBE in light of water contamination and health concerns. Unlike ethanol, MTBE is highly soluble in water and travels easily and swiftly to ground and surface water supplies. Even small amount of methanol either swallowed or absorbed through the skin is very harmful. It can cause blindness, permanent neurological damage, and death. The potential health hazards from the use of MTBE were also documented in a report from the University of California-Davis titled "Health and Environmental Assessment of

Table 4.2 Ethanol blends used in Brazil

Year	Ethanol blend
1987–1988	E22
1993–1998	E22
2000	E20
2001	E22
2003	E20-25
2004	E20
2005	E22
2006	E20
2007	E23-25
2008	E25
2009	E25
2010	E20-25
2011	E18-E25

Based on http://en.goldenmap.com/Common_ethanol_fuel_mixtures

MTBE" which concluded that there are significant risks and costs associated with water contamination due to the use of MTBE. Researchers also found MTBE in over 10,000 groundwater sites in California. The report supports the use of ethanol in place of MTBE, stating that the use of ethanol as an oxygenate would result in much lower risk to water supplies, lower water treatment costs in the event of a spill, and lower monitoring costs.

4.1.1 E10 or Less

E10 is sometimes called *gasohol*. It is a fuel mixture of 10 % anhydrous ethanol and 90 % gasoline that can be used in the ICE of most modern automobiles and light-duty vehicles and without need for any modification on the engine or fuel system. E10 blends are typically rated as 2–3 octane higher than regular gasoline and are approved for use in all new US automobiles and are mandated in some areas for emissions and other reasons. The E10 blend and lower ethanol content mixtures have been used in several countries, and its use has been primarily driven by the several world energy crises that have taken place since the 1973 oil crisis (EAIP 2001).

Other common blends include E5 and E7. These concentrations are generally safe for recent engines that run on pure gasoline. As of 2006, mandates for blending bioethanol into vehicle fuels had been enacted in at least 36 states/provinces and 17 countries at the national level, with most mandates requiring a blend of 10–15 % ethanol with gasoline.

E10 and other blends of ethanol are considered to be useful in decreasing US dependence on foreign oil and can reduce CO emissions by 20–30 % under the right conditions. Although E10 does decrease emissions of CO and greenhouse gases such as carbon dioxide by an estimated 2 % over regular gasoline, it can cause increases in evaporative emissions and some pollutants depending on factors like the age of the vehicle and weather conditions. E10 is also commonly available in the Midwestern USA. E10 has also been mandated for use in all standard automobile fuel in the state of Florida by the end of 2010. Due to the phasing out of MTBE as a gasoline additive and mainly due to the mandates established in the Energy Policy Act of 2005 and the Energy Independence and Security Act of 2007, ethanol blends have increased throughout the USA, and by 2009, the ethanol market share in the US gasoline supply reached almost 8 % by volume. The Tesco chain of supermarkets in the UK have started selling an E5 brand of gasoline marketed as 99 RON super unleaded. Its selling price is lower than the other two forms of high-octane unleaded on the market, Shell's V-Power (99 RON) and BP's Ultimate (97 RON). Many petrol stations throughout Australia now also sell E10, typically at a few cents cheaper per liter than regular unleaded.

In October 2010, the EPA granted a waiver to allow up to 15 % of ethanol blended with gasoline to be sold only for cars and light pickup trucks with a model year of 2007 or later, representing about 15 % of vehicles on the US roads. In January 2011, the waiver was expanded to authorize use of E15 to include model year 2001 through 2006 passenger vehicles. The EPA also decided not to grant any waiver for E15 use in any motorcycles, heavy-duty vehicles, or non-road engines because current testing data do not support such a waiver. According to the Renewable Fuels Association, the E15 waivers now cover 62 % of vehicles on the road in the USA, and the ethanol group estimates that if all 2001 and newer cars and pickups were to use E15, the theoretical blend wall for ethanol use would be approximately 17.5 billion gallons per year. EPA is still studying if older cars can withstand a 15 % ethanol blend.

4.1.2 E20, E25

These blends have been widely used in Brazil since the late 1970s. E20 contains 20 % ethanol and 80 % gasoline, while E25 contains 25 % ethanol. As a response to the 1973 oil crisis, the Brazilian government made mandatory the blend of ethanol fuel with gasoline, fluctuating between 10 and 22 % from 1976 until 1992. Due to this mandatory minimum gasoline blend, pure gasoline is no longer sold in Brazil. A federal law was passed in October 1993, establishing a mandatory blend of 22 % anhydrous ethanol (E22) in the entire country. This law also authorized the executive to set different percentages of ethanol within pre-established boundaries, and since 2003, these limits were fixed at a maximum of 25 % (E25) and a minimum of 20 % (E20) by volume. Since then, the government has set the percentage on the ethanol blend according to the results of the sugarcane harvest and ethanol production from sugarcane, resulting in blend variations even within the same year.

The mandatory blend was set at 25 % of anhydrous ethanol (E25) by Executive Decree since July 1, 2007, and this has been the standard gasoline blend sold throughout Brazil most of the time as of 2011. All Brazilian automakers have adapted their gasoline engines to run smoothly with this range of mixtures. Some vehicles might work properly with lower concentrations of ethanol; however, with a few exceptions, they are unable to run smoothly with pure gasoline which causes engine knocking, as vehicles traveling to neighboring South America countries have demonstrated. Flexible-fuel vehicles, which can run on any mixed of gasoline E20-E25 up to 100 % hydrous ethanol (E100 or hydrated ethanol) ratios, were first available in mid-2003. In July 2008, 86 % of all new light vehicles sold in Brazil were flexible fuel, and only two carmakers build models with a flex-fuel engine optimized to operate with pure gasoline. A state law approved in Minnesota in 2005 mandates that ethanol comprises 20 % of all gasoline sold in this American state beginning in 2013. Successful tests have been conducted to determine the performance under E20 by current vehicles and fuel dispensing equipment designed for E10.

4.1.3 E70, E75

When the vapor pressure drops below 45 kPa in the ethanol blend, fuel ignition cannot be assured on cold winter days. This limits the maximum ethanol blend percentage during the winter months to E75. E70 and E75 are the winter blends used in the United States and Sweden for E85 flexible-fuel vehicles during the cold weather, but still sold at the pump labeled as E85. The seasonal reduction in the ethanol content to an E85 winter blend is mandated to avoid cold-start problems at low temperatures. In the USA, this seasonal reduction in the ethanol content to E70 applies only in cold regions, where temperatures fall below 32 °F during the winter. In Sweden, all E85 flexible-fuel vehicles use an E75 winter blend. This blend was introduced since the winter 2006–2007, and E75 is used from November until March. All E85 flex vehicles require an engine block heater to avoid cold-starting problems. The use of this device is also recommended for gasoline vehicles when temperatures drop below −23 °F. Another option when extreme cold weather is expected is to add more pure gasoline in the tank, thus reducing the ethanol content below the E70 winter blend, or simply not to use E85 during extreme low temperature.

4.1.4 E85

E85 is generally the highest ethanol fuel mixture found in USA and Europe, as this blend is the standard fuel for flexible-fuel vehicles. This mixture has an octane rating of about 105, which is significantly lower than pure ethanol but still higher than normal gasoline (87–95 octane, depending on country). The 85 % limit in the ethanol content was set to reduce ethanol emissions at low temperatures and to avoid cold-starting problems during cold weather, at temperatures lower than 11 °C. A further reduction in the ethanol content is used during the winter in regions where temperatures fall below 0 °C, and this blend is called Winter E85, as the fuel is still sold under the E85 label. A winter blend of E70 is mandated in some regions in the USA, while Sweden mandates E75. As of October 2010, there were nearly 3,000 E85 fuel pumps in Europe, led by Sweden with 1,699 filling stations. The United States had 2,414 public E85 fuel pumps located in 1,701 cities by October 2010, mostly concentrated in the Midwest.

4.1.5 E100

E100 is pure ethanol fuel. Hydrous ethanol as an automotive fuel has been widely used in Brazil since the late 1970s for neat ethanol vehicles and more recently for flexible-fuel vehicles. The ethanol fuel used in Brazil is distilled close to the azeotrope mixture of 95.63 % ethanol and 4.37 % water which is approximately 3.5 % water by volume. The azeotrope is the highest concentration of ethanol that can be achieved via distillation. The maximum water concentration according to the ANP specification is 4.9 vol.%. The E nomenclature is not adopted in Brazil, but hydrated ethanol can be tagged as E100 meaning that it does not have any gasoline, because the water content is not an additive but rather a residue from the distillation process. However, straight hydrous ethanol is also called E95 by some authors. The first commercial vehicle capable of running on pure ethanol was the Ford Model T, produced from 1908 through 1927. It was fitted with a carburetor with adjustable jetting, allowing use of gasoline or ethanol, or a combination of both. At that time, other car manufactures also provided engines for ethanol fuel use. Thereafter, and as a response to the energy crisis, the first modern vehicle capable of running with pure hydrous ethanol (E100) was launched in the Brazilian market, the Fiat 147 after testing with several prototypes developed by the Brazilian subsidiaries of Fiat, Volkswagen, General Motors, and Ford. Since 2003, Brazilian newer flexible-fuel vehicles are capable of running on pure hydrous ethanol ethanol (E100) or blended with any combination of E20–E25 gasoline (a mixture made with anhydrous ethanol), the national mandatory blend.

E100 imposes a limitation on normal vehicle operation as ethanol's lower evaporative pressure (as compared to gasoline) causes problems when cold starting the engine at temperatures below 15 °C. For this reason, both pure ethanol and E100 flexible-fuel vehicles are built with an additional small gasoline reservoir inside the engine compartment to help in starting the engine when cold by initially injecting gasoline. Once started, the engine is then switched back to ethanol. An improved flex engine generation was developed to eliminate the need for the secondary gas tank by warming the ethanol fuel during starting and allowing flex vehicles to do a normal cold start at temperatures as low as −5 °C the lowest temperature expected anywhere in the Brazilian territory. The Polo E-Flex, launched in March 2009, was the first flex-fuel model without an auxiliary tank for cold start. The higher fuel efficiency of E100 in high-performance race cars resulted in Indianapolis 500 races in 2007 and 2008 being run on 100 % fuel-grade ethanol.

4.1.6 Niche Markets

There are also many smaller niche markets where ethanol can be utilized. They are considered niche markets because they are much smaller than the general transportation fuel market. However, they can still create a considerable demand, especially for a much smaller fuel-producing industry like ethanol. Opportunities exist as ethanol-blended diesel fuel, as a hydrogen source for fuel cells, as an aviation fuel, snowmobiles and other off-road vehicles, boats and other personal watercraft, and small engine equipment (RFA 2004, 2005, 2006a, b, c).

4.1.7 Fuel Cells

One of the newest markets being looked at for ethanol use is fuel cells. Fuel cells create electricity by combining hydrogen and oxygen. Fuel cells are more energy efficient than the internal combustion engine. Although fuel cells are considered a new technology for vehicles and other applications, they have actually been around for quite awhile. In fact, the basic configuration and idea for fuel cells was discovered in 1839. However, the commercial possibilities of fuel cells were not realized till the 1960s when NASA began using fuel cells in spacecraft to provide power. In 1998, Chicago became the first city to use pure hydrogen-powered fuel cell buses. Currently, all big three automobile manufacturers, as well as many foreign auto manufacturers, are developing fuel cell vehicle prototypes. Fuel cells could operate on hydrogen from a number of sources including ethanol, methanol, gasoline, and perhaps, at some point, even water. Ethanol is expected to play an important role as a fuel cell fuel supplying the hydrogen for fuel cell operation (RFA 2005). Ethanol is a clean, renewable, hydrogen source that provides huge benefits in reducing the greenhouse gases (GHG) that contribute to climate change (Fuel cells 2000). Nuvera fuel cells, formerly Epyx, have developed a fuel cell processor capable of converting ethanol, methanol, and gasoline to hydrogen. They have found that ethanol provides higher efficiencies, fewer emissions, and better performance than other fuel sources, including gasoline. Ethanol is also much less corrosive then methanol. Ethanol use in fuel cells is a great opportunity for consumers, the environment, and the ethanol industry (RFA 2000a, b, c). The RFA has a Fuel Cell Task Force that continues to follow fuel cell developments and to position ethanol for a role in fuel cell. In USA, the "Hydrogen Fuel Initiative" supports research and commercialization of fuel cells for automobiles and power generation. The first commercial demonstration of ethanol's potential to produce hydrogen to power a fuel cell is underway at Aventine Renewable Energy, Inc.'s Pekin, Illinois ethanol plant. The 13-kilowatt stationary fuel cell system is generating power for the plant's visitor center and additional energy for the plant. The project is a partnership with the US Department of Energy, Caterpillar, Inc., Nuvera fuel cells, the State of Illinois, Renewable Fuels Association and Illinois Corn Growers Association.

Fuel cells work very similarly to batteries except they can run continuously as long as fuel is supplied (Fuel Cells 2000). Benefits of fuel cell use in the transportation sector include increased energy efficiency, a tremendous decrease in emissions, less vehicle maintenance, and the ability to achieve up to 80 mpg. Fuel cells create electricity by combining hydrogen and oxygen. They could eventually be used to supply power to homes, vehicles, and small electronic devices.

Two major hurdles in the development of fuel cells for the transportation sector are storing hydrogen onboard a vehicle and developing a hydrogen refueling infrastructure. One way around these obstacles is to create a process to extract hydrogen from already established fuels such as gasoline and ethanol. Benefits of using ethanol versus other fuels include lower emissions and higher efficiency. However, extracting hydrogen from ethanol requires higher temperatures than when using other fuels, such as methanol, and some argue that this may hinder ethanol use in fuel cells. Ethanol may make up for this shortcoming by increasing the amount of energy over methanol (25 % higher), which allows for a longer vehicle range. Ethanol is also much less corrosive then methanol.

4.1.8 E Diesel

Another market for ethanol considered to be in the developmental stage is the use of ethanol blends in diesel fuel (Launder 1999). These blends range from 7.7 to 15 % ethanol and 1 to 5 % special additives that prevent the ethanol and diesel from separating at very low temperatures or if water contamination occurs. Ethanol-blended diesel could provide a considerable increase in demand for ethanol as diesel vehicles in the USA consume approximately 50 billion gallons a year. Demonstrations are currently being conducted on the use of ethanol-blended diesel in heavy-duty trucks, buses, and farm machinery.

The use of ethanol-blended diesel has similar disadvantages to other ethanol-blended fuel, including decrease in gas mileage and increase in cost compared to regular diesel. Advantages include a decrease in emissions and in demand for imported petroleum. Presently, E diesel fuels are considered experimental and are being developed by many companies. Blends containing up to 15 vol.% ethanol, blended with standard diesel and a proprietary additive, are called E diesel fuels. A number of fleet demonstrations have been completed with favorable results, and some controlled testing has also been completed. Results show that E diesel blends reduce certain exhaust emissions, especially particulates, in certain diesel applications and duty cycles. However, additional testing is needed to assess compatibility of E diesel blends with various fuel system parts and also to determine the long-range effects on engine durability. Additional emissions tests are also needed to more accurately quantify the emission profiles of various E diesel blend levels. The RFA is working closely with industry to address the research needs (RFA 2005; Vaughn 2000). Most E diesel blends have properties similar to standard diesel fuel or can be modified to be similar through the use of additives.

One of the important differences is the lower flash point of E diesel blends. E diesel blends are designated as Class I, whereas diesel fuel is designated as a Class II flammable liquid.

4.1.9 Aviation

A more near-term niche market for ethanol is aviation. Aviation fuel only accounts for about 5 % of the total transportation fuel use in the United States; it still represents a sizable market for ethanol consumption (approximately 400 million gallons per year). Currently, ethanol is used in some small engine crafts and by experimental aircraft pilots (Caddet 1997; Higdon 1997; Bender 2000). Research on the use of ethanol in aviation started more than 20 years ago. In March 2000, FAA certified AGE-85 (aviation-grade ethanol) which is a high-performance fuel that may be used in any piston engine aircraft. It contains approximately 85 % ethanol, along with light hydrocarbons and biodiesel fuel. AGE-85 is specifically blended for cold starting and good mixture balance. AGE is unleaded, burns cleaner, has lower exhaust emissions, and is more environmentally friendly than traditional aviation fuels. The ethanol in AGE-85 prevents carburetor and fuel line icing and provides excellent detonation margins. Some constraints in using AGE-85 include its current lack of availability and decreased range (5–15 %) due to its lower BTU content. Probably, the most formidable constraint to ethanol being excepted in aviation is some opposition within the field itself. The Aircraft Owners and Pilots Association (AOPA) and Cessna have stated that any new unleaded fuel being considered for aircraft should be able to be used without aircraft modifications. They also are doubtful about the ability and willingness of airports to put up a new refueling infrastructure to accommodate a fuel such as AGE-85. Indeed, pilots would most likely have to secure funding or put up the money themselves to install E-85 refueling equipment—at least until widespread use significantly increases demand (Launder 1999).

There is also the potential for ethanol to be used in a biodiesel blend and/ or ETBE in turbine aircraft, which would not require aircraft modifications. A biodiesel blend would be 20 % biodiesel–80 % regular diesel and could reduce particulate emissions by 80 %. ETBE contains about 43 % ethanol and isobutylene a gas/petroleum derivative. This could also prove to be a considerable market for ethanol as turbine aircraft used approximately 16.4 billion gallons of fuel in 1997.

Presently, ethanol is used in some small engine crafts and by experimental aircraft pilots. The largest general aviation market in the world is the USA, and the second largest is Brazil. Pressure is being put by EPA on the aviation sector to take the lead out of their fuel. It appears likely that Avgas (the leaded aviation standard fuel) will need to be replaced with an unleaded fuel in the near future and that ethanol could be the replacement fuel.

4.1.10 Snowmobiles

Another potential market for the use of ethanol is snowmobiles. Snowmobiles could provide a considerable market for ethanol in USA. There are a total of over 2.3 million snowmobiles registered in the United States. The snowmobile manufacturers approve the use of ethanol blends of up to 10 %. The use of ethanol in snowmobiles greatly reduces emissions. In the USA, several million snowmobiles are registered. This is a significant factor to consider as a snowmobile with a conventional two-stroke engine emits 36 times more carbon monoxide and 98 times more hydrocarbons than an automobile. In fact, snowmobiles can create serious environmental and health risks that many national parks are considering limiting or banning the use of snowmobiles within the parks. Using a blend of 10 % ethanol also eliminates the need for the use of gasoline antifreeze and removes and prevents deposit build up in the fuel tank (Launder 1999).

4.1.11 Boats/Marine

Ethanol has been shown to be a viable alternative to gasoline for use in recreational boat engines, due to better environmental performance as a fuel over gasoline (Launder 1999). Many marine/boat manufacturers approve the use of ethanol blends of up to 10 %. Study conducted by Dambach et al. (2004) has proven the ability to retrofit an engine with minimal modifications and lose little in the way of performance. The use of alcohol-based fuels in older boats may cause some problems as the older boats may not have alcohol compatible parts, and therefore, certain parts could be degraded with the use of ethanol-blended fuels. Ethanol blends of up to 10 % can also be used in boats and personal watercraft. This could also prove to be a considerable size market USA. Many marine/boat manufacturers approve the use of ethanol-blended fuels. However, some boat manufacturers do have cautionary language on the use of alcohol-based fuels in older boats as they may not have alcohol compatible parts, and therefore, certain parts could be degraded with the use of ethanol-blended fuels.

4.1.12 Small Engine Equipment

Ethanol blends can also be used in small engines such as lawn mowers, chainsaws, and weed trimmers (Launder 1999). The benefits of using ethanol would again be decreased emissions. The Portable Power Equipment Manufacturers Association has conducted reformulated fuel research for chainsaws, weed trimmers, and other handheld equipment and observed no operating problems with equipment when using reformulated gasoline. Study conducted by California Environmental

Protection Agency has shown that a gasoline-powered lawn mower run for an hour puts out about the same amount of smog-forming emissions as 40 new automobiles run for an hour. They also found that a chain saw operated for 2 h is able to generate the same amount of emissions as ten automobiles driven 250 miles.

References

ANFAVEA—National Association of Automobile Fabricants (2006) Brazilian automotive industry yearbook. p 167 (in Portuguese)

Bajpai P (2007) Bioethanol. PIRA Technology Report, Smithers PIRA, UK

Bender B (2000) Ethanol: aviation fuel of the future. Michigan Ethanol Workshop Presentation

Caddet (1997) Ethanol as an aviation fuel. Caddet renewable energy. Technical Brochure No. 51. Available at http://lib.kier.re.kr/caddet/retb/no51.pdf

Clean Fuels Development Coalition (CFDC) (1999) Fuel ethanol fact book. Bethesda

Dambach E, Han A, Henthorn B (2004) Ethanol as fuel for recreational boats. ENGS 190/ENGG 290 final report. Available at http://www.dartmouth.edu/~ethanolboat/Ethanol_Outboard_Final_Report.pdf

EAIP (Earth Policy Institute) (2001) Setting the ethanol limit in petrol. Environment Australia issues paper, January

Fuel Cells (2000) Benefits of fuel cells. Available at http://216.51.18.233/fcbenefi.html

Higdon D (1997) Treaty calls for Great Lakes ban on leaded fuel. General Aviation News & Flyer

IEA (International Energy Agency) (2004) Biofuels for transport—an International perspective. Paris. Available at http://www.iea.org/textbase/nppdf/free/2004/biofuels2004.pdf

IEA (International Energy Agency) (2006) Medium term oil market report. Paris. Available at http://www.iea.org

Launder K (1999) Opportunities and constraints for ethanol-based transportation fuels. Lansing: State of Michigan, Department of Consumer and Industry Services, Biomass Energy Program. Available at http://www.michigan.gov/cis/0,1607,7-154-25676_25753_30083-141676-,00.html

RFA—Renewable Fuels Association (2004) Ethanol industry outlook: synergy in energy. Available at http://www.ethanolrfa.org/objects/pdf/outlook/outlook_2004.pdf

RFA—Renewable Fuels Association (2005) Ethanol industry outlook: homegrown for the homeland. Available at http://www.ethanolrfa.org/objects/pdf/outlook/outlook_2005.pdf

RFA—Renewable Fuels Association (2006a) Ethanol industry outlook: from niche to nation. Available at www.ethanolrfa.org/objects/pdf/outlook/outlook_2006.pdf

RFA—Renewable Fuels Association (2006b) Fuel ethanol industry plants and production capacity. U.S

RFA—Renewable Fuels Association (2006c) Statistics data. Available at http://www.ethanolrfa.org/industry/statistics/#C

RFA—Renewable Fuels Association (2000a) Ethanol as a renewable fuel source for fuel cells. Available at http://www.ethanolRFA.org/fuelcells.htm

RFA—Renewable Fuels Association (RFA) (2000b) President issues executive order to green the federal fleet on Earth Day. Ethanol Report. 27 April 2000

RFA—Renewable Fuels Association (RFA) (2000c) Ethanol—its use in gasoline: expected impacts and comments of expert reviewers. http://www.ethanolrfa.org/objects/documents/87/ethanol_tox_20001.pdf

Vaughn E (2000) RFA. International fuel ethanol workshop presentation. Available at www.sentex.net/~crfa/crfanew.html—40k

Chapter 5
Characteristics of Fuel Ethanol

Ethanol is more reactive than hydrocarbon fuels, such as gasoline. Since it is an alcohol, its molecular structure shows a polar fraction due to the hydroxyl radical and a nonpolar fraction in its carbon chain. That explains why ethanol can be dissolved in both gasoline (nonpolar) and in water (polar). Due to its short carbon chain, the properties of ethanol polar fraction overcome the nonpolar properties. The formation of hydrogen bridges in ethanol molecule results in higher boiling temperature in comparison to that of gasoline (Table 5.1). Ethanol is less toxic than methanol—another alcohol used as fuel. The simple structure of ethanol molecule makes it suitable for spark ignition internal combustion engines operation. The high octane number of ethanol allows for higher compression ratios in comparison to gasoline fueled engines (Table 5.1) (Costa and Sodre 2010). One of the main advantages of ethanol offers when compared to gasoline is its antiknock performance, allowing its use in higher compression ratio engines. At high temperature, ethanol produces superior thermal efficiency due to its higher heat of vaporization. As ethanol can burn richer fuel/air mixtures, it allows for higher engine power output in comparison to gasoline. However, due to its lower heating value, the use of ethanol instead of gasoline results in higher fuel consumption. Engine cold start is also a problem for ethanol, due to its low vapor pressure (Owen and Coley 1995). Rasskazchikova et al. (2004) discussed the use of ethanol as high-octane additive to automotive gasoline. The authors concluded that, despite the high cost, ethanol is the most promising octane-raising additive available. That is justified by the low toxicity, reduced environmental pressure when burning ethanol-containing fuel, and production from renewable raw material. Silva et al. (2005) evaluated the effect of ethanol and other additives on the antiknock properties and Reid vapor pressure (RVP) of gasoline. Addition of ethanol up to 25 % by volume in gasoline led to increased RVP and octane ratings.

Ethanol is used as an automotive fuel and can be used alone in specially designed engines, or blended with gasoline and used without any engine modifications.

Some excerpts taken from Bajpai (2007). PIRA Technology Report on Bioethanol with kind permission from Smithers PIRA

Table 5.1 Properties of hydrous ethanol blend and gasoline–ethanol blend

Parameter	Hydrous ethanol	78 % Gasoline 22 % ethanol
Chemical structure	C2 H6.16 O1.08	C6.39 H13.60 O0.61
Carbon mass (%)	50.59	76.7
Hydrogen mass (%)	12.98	13.6
Oxygen mass (%)	36.42	9.7
Sulfur mass (%)	0	0.09
Self-ignition temperature (°C)	420	400
Temperature of vaporization (°C)	78	40–220
Heat of vaporization (kcal/kg)	105	237
Research octane number (RON)	–	106
Motor octane number (MON)	87	80
Vapor pressure (bar)	29	27.5
Laminar flame speed (m/s)	0.42	0.30
Density (kg/l) 0.74 0.81	0.81	0.74
Lower heating value (kcal/kg)	5970	9400
Stoichiometric air/fuel ratio	8.7	13.1

Based on Costa and Sodre (2010)

Motorboats, motorcycles, lawn mowers, chain saws, etc. can all utilize the cleaner gasoline/ethanol fuel. Ethanol is a high-octane fuel with high oxygen content. So, it allows the engine to more completely combust the fuel, resulting in fewer emissions. Ethanol has many advantages as an automotive fuel. Ethanol blends also reduces carbon monoxide emissions. When used in a correctly formulated fuel, ethanol can also reduce vehicle emissions which contribute to the formation of smog. It can also withstand cooler temperatures. Because of its low freezing point, it can be used as the fluid in thermometers for temperatures below −40 °C, and for other low temperature purposes, such as for antifreeze in automobile radiators.

The Energy Policy Act of 2005 established RFS, which mandates the use of ethanol and other renewable fuels in gasoline. Most of the fuel ethanol consumed in USA is E10. A blend of 85 % ethanol and 15 % unleaded gasoline (E85) is also being used. Currently, in USA, approximately 50 million gallons of ethanol are made into E85. E85 is available at 1,000 locations in the USA (including both public and private). The National Ethanol Vehicle Coalition estimates approximately 6 million FFVs on America's roads today as compared to approximately 230 million gasoline- and diesel-fueled vehicles. Most E85-capable vehicles are "flexible fuel vehicles" or FFVs (RFA 1999, 2001, 2006a, b, c, 2007a, b). This mixture has an octane rating of about 105. This is significantly lower from pure ethanol but still much higher than normal gasoline. The addition of a small amount of gasoline helps a conventional engine start when using this fuel under cold conditions. E85 does not always contain exactly 85 % ethanol. In winter, particularly in colder climates, additional gasoline is added to facilitate cold start. E85 has been similar in cost to gasoline, but with the large oil price rises of 2005, it is being sold for as much as $0.70 less per gallon than gasoline which makes it highly attractive to the small but growing number of motorists with cars capable of burning it. Gasoline contains more energy, gallon for gallon than ethanol.

One gallon of gasoline contains approximately 114,132 btu and ethanol contains 76,000 btu. Therefore, E85 contains approximately 27 % less btu then 100 % gasoline. However, when factoring in the fuel efficiency of gasoline and ethanol (which is more fuel efficient), it was found that E85 mileage reduction is only about 10–15 % instead of the 27 % based strictly on energy content.

E85 is environmentally friendly (Niven 2005). It has the highest oxygen content of any fuel available today, making it burn cleaner than ordinary gasoline. The use of E85 reduces pollutants such as ozone and carbon monoxide and air toxins like benzene. E85 cars perform well with significant reductions in emissions when compared to vehicles using ordinary unleaded gasoline. Reductions in two particularly troublesome pollutants, carbon monoxide and hydrocarbons, are reduced significantly. Ethanol is one of only two liquid fuels available that combats global warming because of its raw material source. As corn grows, it converts carbon dioxide into oxygen. Auto makers are offering more flexible fuel vehicles. Purchase price of these vehicles has been comparable to the base price of gasoline models. Since E85 is a cleaner burning fuel, it is expected that the life of a flexible fuel vehicle will be somewhat longer than that of a comparable gasoline vehicle.

The key component in a variable fuel vehicle is a sensor that determines the percentage of ethanol in the fuel. With the help of a computer, the vehicle automatically adjusts for best performance and emissions. Chrysler began offering E85 minivans in the 1998 model year, and Ford offered the Taurus and added Windstar and Ranger to the E85 flexible fuel vehicles in the 1999 model year. Ford, GMC, Chevrolet, and Daimler-Chrysler are now offering E85 variable fuel vehicles.

The mileage decrease which occurs when operating a vehicle on E85 has been debated. Ford reports an average of 16 MPG for E85 Taurus (based on city and highway driving) and 22 MPG for gasoline (Ford Motor Company 1998). Daimler-Chrysler reports a 27 % range reduction when using E85 in their minivans (Chrysler Corporation 1997).

The color of ethanol fuel blends depends on the color of the gasoline in the blend. Blends may also have a gasoline-like odor. A gallon of E85 contains 27 % less energy. The energy content of a gallon of ethanol is equal to 76,000 BTU as compared to about 115,000 BTU for a gallon of conventional gasoline. It follows that to replace the energy equivalent of a gallon of gasoline, we would need approximately 1.5 gallons of ethanol.

The average vehicle on the road today emits more than 600 pounds of pollution into the air each year in US. These harmful emissions include carbon monoxide, volatile organic compounds, particulate matter, oxides of nitrogen, and carbon dioxide. These emissions have significant health implications because they contribute to the amount of smog and carbon monoxide in air. Carbon monoxide emissions have also been implicated in global warming. One of the benefits of using E85 vehicles is a reduction in the amount of pollutants emitted into the air we breathe. The emissions control systems found on ethanol-powered vehicles manufactured today have been engineered to meet or exceed all federal and state emissions control regulations. Two types of emissions are released by E85 vehicles—exhaust and evaporative. Although compliance with federal and state regulations has already resulted in a decrease in exhaust emissions from gasoline-powered vehicles, ethanol-fueled vehicles can further reduce pollution from emissions by a modest amount. Most ethanol-fueled vehicles produce lower carbon monoxide and carbon dioxide emissions and the same or lower levels of hydrocarbon and non-methane hydrocarbon emissions compared with gasoline-fueled vehicles. Nitrogen oxide emissions are about the same for ethanol and gasoline vehicles. Emissions resulting from fuel evaporation are a potential problem for any vehicle, regardless of the fuel. E85 has fewer highly volatile components than gasoline and so has fewer emissions resulting from evaporation.

Brazil has used ethanol blends since 1939. High oil prices in the 1970s prompted a government mandate in Brazil to produce vehicles fueled by pure ethanol in order to reduce dependence on foreign oil and provide value-added markets for its sugar cane producers. Today, there are several million ethanol-powered vehicles in Brazil that consume more than 4 billion gallons of ethanol annually.

Requirements in the Clean Air Act to make cleaner burning reformulated gasoline with lower volatility and fewer toxic components have increased interest in ethanol-based ethers such as ethyl tertiary butyl ether. It is a chemical compound produced by reacting ethanol and isobutylene. ETBE has superior physical and combustion characteristics to other ethers. They include low volatility, high octane value, lower carbon monoxide and hydrocarbon emissions, and superior driveability. Ethanol and ETBE are among the oxygenates used in reformulated gasoline that is required in certain ozone non-attainment areas in the US.

5.1 Ethanol Based Engines

Ethanol is most commonly used to power automobiles, though it may be used to power other vehicles, such as farm tractors and airplanes. Ethanol (E100) consumption in an engine is approximately 34 % higher than that of gasoline (the energy per volume unit is 34 % lower). However, higher compression ratios in an ethanol-only engine allow for increased power output and better fuel economy than would be obtained with the lower compression ratio. In general, ethanol-only engines are tuned to give slightly better power and torque output to gasoline-powered engines. In flexible fuel vehicles, the lower compression ratio requires tunings that give the same output when using either gasoline or hydrated ethanol. For maximum use of ethanol's benefits, a much higher compression ratio should be used, which would render that engine unsuitable for gasoline usage. When ethanol fuel availability allows high compression ethanol-only vehicles to be practical, the fuel efficiency of such engines should be equal or greater than current gasoline engines. However, since the energy content (by volume) of ethanol fuel is less than gasoline, a larger volume of ethanol fuel would still be required to produce the same amount of energy.

A 2004 study and a paper published by the Society of Automotive Engineers present the possibility of a definite advance over hybrid electric cars' cost-efficiency by using a high-output turbocharger in combination with continuous dual fuel direct injection of pure alcohol and pure gasoline in any ratio up to 100 % of either. Each fuel is stored separately, probably with a much smaller tank for alcohol, the peak cost-efficiency being calculated to occur at approximately 30 % alcohol mix, at maximum engine power. The estimated cost advantage is calculated at 4.6:1 return on the cost of alcohol used, in gasoline costs saved, when the alcohol is used primarily as an octane modifier and is otherwise conserved. With the cost of new equipment factored in the data gives a 3:1 improvement in payback over hybrid and 4:1 over turbo-diesel (comparing consumer investment yield only). In addition, the danger of water absorption into premixed gasoline and supply issues of multiple mix ratios can be addressed by this system.

All manufacturers recommend the use of ethanol for environmental reasons. A survey revealed that nine out of ten auto dealers use ethanol-blended gasoline in their personal vehicles. Over half of the dealerships surveyed indicated their customers reported benefits that included reduced knocking and pinging, improved gas mileage, better acceleration, and improved starting qualities. A survey conducted at Iowa indicated that 9 out of 10 technicians used ethanol in their personal vehicles and reported the same benefits as the auto dealers. E10 unleaded is approved under the warranties of all domestic and foreign automobile manufacturers marketing vehicles in the United States. The top three automakers in USA, Daimler-Chrysler, Ford, and General Motors, recommend the use of oxygenated fuels such as ethanol blends because of their clean air benefits and performance qualities. Ethanol is a good cleaning agent. It helps keep the engine clean in newer vehicles. In older vehicles, it can sometimes loosen contaminants and residues that have already been deposited in a vehicle's fuel delivery system. Occasionally, these loosened materials collect in the fuel filter and can then be removed simply by changing the fuel filter. All alcohols have the ability to absorb water. Condensation of water in the fuel system is absorbed and does not have the opportunity to collect and freeze. Since ethanol blends contain at least 10 % ethanol, they are able to absorb water and eliminate the need for adding a gas-line antifreeze in winter. Ethanol is a fuel for old and new engine technology. Automotive engines older than 1969 with non-hardened valve seats may need a lead substitute added to gasoline or ethanol blends to prevent premature valve seat wear. Valve burning is decreased when ethanol blends are used because ethanol burns cooler than ordinary unleaded gasoline. Many high performance racing engines use pure alcohol for that reason.

Modern computerized vehicles perform better than non-computer equipped vehicles when operating correctly. The improvement in performance is due to the vehicle's computerized fuel system being able to make adjustments with changes in operating conditions or fuel type. Some of the chemicals used to manufacture gasoline, such as olefins, have been identified as a cause of deposits on port fuel injectors. Today's gasolines contain detergent additives that are designed to prevent fuel injector and valve deposits.

The quality of fuel used in any motor vehicle engine is very important to its long life and proper operation. Driveability will suffer if the fuel is not right for the air temperature or if fuel changes to a vapor incorrect. Gasoline is a complex mixture of approximately 300 various ingredients, mainly hydrocarbons, refined from crude petroleum oil for use as fuel in engines. Refiners must meet gasoline standards set by the American Society for Testing and Materials (ASTM), the Environmental Protection Agency (EPA), state regulatory agencies, and their own company standards.

MTBE is a popular oxygenate that competes with ethanol for market share as an additive. It is the most widely used fuel additive to make reformulated gasoline; however, it cannot be used as the dominant ingredient in a fuel. MTBE is highly corrosive, more volatile than ethanol, and more damaging to plastic and rubber fuel system components known as elastomers. Both MTBE and ETBE are high octane, low volatility, oxygenated fuel components made by combining alcohol with isobutlylene. MTBE is permitted in unleaded gasoline up to a level of 15 % whereas ETBE can be added to gasoline up to a level of about 17 %. Many car company warranties do not cover the use of methanol-based fuels, while all auto makers approve the use of ethanol-blended gasoline. MTBE has been found to contaminate ground water, and a ban on MTBE has been implemented in half the states in the USA with a reasonable possibility of the ban spreading to the remaining twenty-five states in the near future. As long as a reformulated fuel mandate stays in place, ethanol is virtually guaranteed an expanded market. If the MTBE ban passes, then there is not enough ethanol production capacity to immediately meet excess demand caused by the ban. Therefore, more ethanol infrastructure would need to be built to accommodate the extra demand. A more gradual phaseout would help the ethanol industry adapt to the growing need for its product.

5.2 Fuel Economy

All vehicles have a fuel economy (measured as miles per US gallon -MPG-, or liters per 100 km) that is directly proportional to energy content. Ethanol contains approx. 34 % less energy per unit volume than gasoline and therefore will result in a 34 % reduction in miles per US gallon. For E10, the effect is small (~3 %) when compared to conventional gasoline and even smaller (1–2 %) when compared to oxygenated and reformulated blends. However, for E85, the effect becomes significant. E85 will produce lower mileage than gasoline and will require more frequent refueling. Actual performance may vary depending on the vehicle. This reduced fuel economy should be considered when making price comparisons. For example, if regular gasoline costs $3.00 per gallon, and E85 costs $2.19 per gallon, the prices are essentially equivalent. If the discount for E85 is less than 27 %, it actually costs more per mile to use.

Research is underway to increase fuel efficiency by optimizing engines for ethanol-based fuels. Ethanol's higher octane allows an increase of an engine's compression ratio for increased thermal efficiency. In one study, complex engine controls and increased exhaust gas recirculation allowed a compression ratio of 19.5 with fuels ranging from neat ethanol to E50. Thermal efficiency up to approximately that for a diesel was achieved. This would result in the miles per gallon of a dedicated ethanol vehicle to be about the same as one burning gasoline. There are currently no commercially available vehicles that make significant use of ethanol-optimizing technologies, but this may change in the future.

References

Bajpai P (2007) Bioethanol. PIRA Technology Report, Smithers PIRA, UK

Chrysler Corporation (1997) Chrysler corporation's flexible fuel minivans

Costa RC, Sodre JR (2010) Hydrous ethanol vs. gasoline-ethanol blend: Engine performance and emissions. Fuel 89(2010):287–293

Ford Motor Company (1998) The 1998 E85 ford taurus flexible fuel vehicle

Niven RK (2005) Ethanol in gasoline: environmental impacts and sustainability review article. Renew Sustain Energy Rev 9:535–555

Owen K, Coley T (1995) Automotive fuels reference book, 2nd edn. Society of Automotive Engineers, Inc, USA

Rasskazchikova TV, Kapustin VM, Karpov SA (2004) Ethanol as high-octane additive to automotive gasolines. Production and use in Russia and abroad. Chem Technol Fuels Oils 40(4):203–10

RFA—Renewable Fuels Association (1999) Ford coupon program to encourage use of E85. Ethanol report

RFA—Renewable Fuels Association (2001) Ethanol: clean air, clean water, clean fuel—industry outlook 2001. Available at: http://www.ethanolrfa.org/RFAannualreport01.pdf

RFA—Renewable Fuels Association (2006a) Ethanol industry outlook: from niche to nation. Available at: www.ethanolrfa.org/objects/pdf/outlook/outlook_2006.pdf

RFA—Renewable Fuels Association (2006b) Fuel ethanol industry plants and production capacity. U.S

RFA—Renewable Fuels Association (2006c) Statistics data. Available at: http://www.ethanolrfa.org/industry/statistics/#C

RFA—Renewable Fuels Association (2007a) Ethanol industry outlook: building new horizons. Available at: http://www.ethanolrfa.org/objects/pdf/outlook/RFA_Outlook_2007.pdf

RFA—Renewable Fuels Association (2007b) Cellulosic ethanol grants provide much needed boost to fledgling technology. Available at: www.biofuelsjournal.com/articles/RFA

Silva R, Cataluña R, Menezes EW, Samios D, Piatnicki CMS (2005) Effect of additives on the antiknock properties and Reid vapor pressure of gasoline. Fuel 2005(84):951–959

Chapter 6
Benefits and Problems with Ethanol

6.1 Benefits

Ethanol provides several benefits. It is easily biodegraded in the environment and produces much less pollutants in internal combustion engines than petroleum fuels (Green fuels 1998a, b). It has low toxicity and is miscible with water. Many car makers are producing more vehicles with tolerances to burn high %E fuels more efficiently. Thus, the risk posed by ethanol to the environment is significantly lower than that of fuels produced from petroleum, and the demand for ethanol will increase with time as these automobile improvements take place.

Ethanol is miscible with gasoline in any proportion, but is found most commonly as 10 and 85 % ethanol. FFVs can operate on blends of ethanol and gasoline anywhere between 0 and 85 % ethanol. Benefits of ethanol include higher performance; cleaner burning fuel; positive energy balance; currently cheaper than gasoline (after considering subsidies). A notable fact is that fuel ethanol is versatile and can basically be used in two capacities: a fuel additive (E10) and an almost stand-alone fuel (E85). Different issues surround each type of fuel. Any internal combustion engine, including small engines such as those in lawn mowers, can use a blend up to 10 % ethanol. Ethanol has a higher octane rating than most gasoline, which means that engines burning ethanol are less likely to "knock." Pure ethanol has an octane number of 112, while E85 is about 105 octane. Such a high octane rating means that even high performance engines can use ethanol fuels. Ethanol is a good racing fuel because it combusts at a lower temperature than gasoline, thus requiring less cooling power from the radiator.

Ethanol as a fuel has a positive energy balance, which means that combustion of one unit of ethanol yields more energy than the aggregate energy required to produce the inputs that make up that unit. It follows that ethanol is more beneficial for the environment than using only petroleum based fuels.

Some excerpts taken from Bajpai (2007). PIRA Technology Report on Bioethanol with kind permission from Smithers PIRA

P. Bajpai, *Advances in Bioethanol*, SpringerBriefs in Applied Sciences and Technology, DOI: 10.1007/978-81-322-1584-4_6, © The Author(s) 2013

The U.S. Environmental Protection Agency considers ozone to be the most widespread air pollution problem (EPA 1997). To combat this problem, ethanol is widely used in reformulated gasolines to help urban cities meet public health standards for ozone. Because it is produced from renewable resources, ethanol is the only transportation fuel that reduces greenhouse gas emissions from cars. Fossil fuels release carbon trapped in the soil into the air, where it reacts with oxygen to form carbon dioxide, a greenhouse gas that traps the earth's heat, contributing to global warming (EPA 2001).

Ethanol is made from agricultural crops, which take carbon dioxide and give off oxygen. This maintains the balance of carbon dioxide in the atmosphere. Increased use of renewable fuels like ethanol will help counter the pollution and global warming effects of burning gasoline. Under current conditions, use of ethanol blended fuels as E85 can reduce the net emissions of greenhouse gases by as much as 30–36 % and can further contribute by decreasing fossil energy use by 42–48 % (Hu et al. 2004). Ethanol blended fuel as E10 reduces greenhouse gases by 2.4–2.9 % and fossil energy use by 3.3–3.9 %.

6.2 Problems with Ethanol/Ethanol Blends

Ethanol/ethanol blends create problem during storage, transportation, and combustion (Orbital Engine Company 2002a, b, 2003). Main drawback of using ethanol with gasoline is phase separation because it is immiscible with diesel fuel over a wide range of temperatures.

One of the most challenging issues related to ethanol fuel blends involves the stability of mixtures. The shelf life of ethanol fuel blends is much lower due to its water absorbing and corrosive qualities. It does not store longer than 2 or 3 months without adding a stabilizer (US DOE 2006a, b). Even so, Ethanol stored in fiberglass or plastic tanks will make the fiberglass soft and mushy, leading to tank failure and engine failure, because the plastic or fiberglass is dissolving into the gas.

Ethanol cannot travel in pipelines along with gasoline, because it picks up excess water and impurities. As a result, ethanol needs to be transported by trucks, trains, or barges, which is more expensive and complicated than sending it down a pipeline. It would be a much expensive affair to build a completely new pipeline network specifically for ethanol.

Some of the soft metals such as zinc, brass, lead, aluminum, copper are not compatible with ethanol and can suffer corrosion and pitting if exposed to it for extended periods (US DOE 2006b). Furthermore, ethanol dissolves in water and is more electrically conductive than petrol. The presence of water can facilitate corrosion, and the conductivity facilitates the possibility of galvanic corrosion.

Ethanol has a solvent effect and will loosen gums and other deposits in fuel systems that have been operated on mineral petrol a long time. In extreme cases, this can clog fuel filters and cause the engine to run poorly. Degradation of fuel system components can potentially occur because the materials used for hoses, seals, O-rings, membranes, and gaskets are not compatible with ethanol. The typical mechanism is that the ethanol is absorbed into the material and breaks down the molecular bonds within it. This can lead to swelling of the material, softening, or embitterment and eventually failure of the component.

Ethanol blend fuels have an affinity to absorb much greater extent of water, very quickly, compared to conventional non-alcohol gasoline (ethanol gasoline blends can absorb 50 times more water than conventional non-alcoholic gasoline). Some water can be dissolved in ethanol-petrol blends and will pass through the fuel system with no effect. However, if the amount of water present is too great, the blend will separate into an upper petrol layer and a lower water/ethanol layer. Generally, fuel is drawn from the bottom of the tank, so the water/ethanol layer will be drawn into the engine first and the engine will not run (Lapuerta et al. 2007; Bhattacharya and Mishra 2003).

The ethanol molecule contains oxygen while the main components of petrol do not. The effect of this is that less oxygen from air is required to achieve complete combustion, and so if the air/fuel mixture is not adjusted, the mixture is leaner than it would be on pure petrol. This can lead to engine operability problems such as hesitancy at full throttle and/or higher exhaust temperatures (Orbital Engine Company 2002a). The volatility characteristics of an ethanol/petrol blend differ from those of the unblended petrol on its own. These differences in volatility can impact on engine operability.

Ethanol fuel blend is not compatible with some vehicle parts like capacitance fuel level gauging indicators as they give erroneous fuel quantity indications in vehicles. The use of ethanol-based fuels can negatively affect electric fuel pumps by increasing internal wear and undesirable spark generation.

Small scale production of ethanol requires a significant input of equipment and labor. There is added cost to process the alcohol to the required 200 proof for blending with gasoline for unmodified engines. To be most effective, the ethanol should be used in modified engines.

An ethanol and gasoline mixture can cause damage to the environment during transportation, storage, and consumption. Once the mixture is leaked or spilled into surface or groundwater, ethanol will hamper degradation of other toxic components in gasoline. Ethanol also increases water solubility of gasoline, leading to larger contamination areas. Moreover, improvement of air quality is still uncertain. While an ethanol and gasoline mixture has lower sulfuric oxide and carbon dioxide emissions, it has higher nitrogen oxide and VOC emissions, which causes ozone depletion (Jia et al. 2005).

References

Bajpai P (2007) Bioethanol. PIRA Technology Report, Smithers PIRA, UK

Bhattacharya TK, Mishra TN (2003) Studies on feasibility of using lower proof ethanol-diesel blends as fuel for compression Ignition engines. Inst Eng: Agric Eng 84:56–59

Department of Energy (2006a) Bioethanol feedstocks. Available at: http://www1.eere.energy.gov/biomass/abcs_biofuels.html

Department of Energy (2006b) Guidebook for handling, storing, & dispensing fuel ethanol prepared for the U.S. department of energy by the center for transportation research energy systems division, Argonne National Laboratory

EPA (Environmental Protection Agency) (1997) Fact sheet: EPA's revised ozone standard, United States. Available: www.epa.gov/ttn/oarpg/naaqsfin/o3fact.html. Mar 2001

EPA (Environmental Protection Agency) (2001) Fact sheet: global warming: climate, United States. Available: http://www.epa.gov/globalwarming/climate/index.html

Green Fuels (1998a) Environmental effects of ethanol and gasoline. Available at: http://www.greenfuels.org/ethaenv1.html

Green Fuels (1998b) How does ethanol clear the air? Available at: http://www.greenfuels.org/ethaair.html

Hu Z, Pu G, Fang F, Wang C (2004) Economics, environment, and energy life cycle assessment of automobiles fueled by bio-ethanol blends in China. Renew Energy 29:2183–2192

Jia LW, Shen MQ, Wang J, Lin MQ (2005) Influence of ethanol–gasoline blended fuel on emission characteristics from a four-stroke motorcycle engine. J Hazard Mater 123(1–3):29–34

Lapuerta M, Armas O, García R (2007) Stability of diesel–bioethanol blends for use in diesel engines. Fuel 86(10–11):1351–1357

Orbital Engine Company (2002a) A literature review based assessment on the impacts of a 10 and 20 % ethanol gasoline fuel blend on the Australian vehicle fleet. Canberra

Orbital Engine Company (2002b) A literature review based assessment on the impacts of a 20 % ethanol gasoline fuel blend on the Australian vehicle fleet. Canberra

Orbital Engine Company (2003). Market barriers to the uptake of a biofuels study. A testing based assessment to determine impacts of a 20 % ethanol gasoline fuel blend on the Australian passenger vehicle fleet-2,000 hrs material compatibility testing. Canberra

Chapter 7
Global Production of Bioethanol

Several countries have introduced or are introducing programs for fuel ethanol. According to RFA, Brazil and the US are the dominant industrial players, accounting for 87 % of global biofuel production driven by government support. Tables 7.1, 7.2 shows Total world ethanol production and Ethanol Industry Overview. Figure 7.1 shows Global Ethanol Facilities, and Fig. 7.2 shows ethanol production by type. Below is a description about bioethanol programs worldwide:

7.1 European Union

The European Commission is supporting biofuels with the aim of reducing greenhouse gas emissions, boosting the decarbonization of transport fuels, diversifying fuel supply sources, offering new income opportunities in rural areas, and developing long-term replacements for fossil fuel.

In EU, Ethanol is mainly produced from sugar beets and wheat. Sugar beets prove to be a good feedstock for European bioethanol production. Because sugar beets have a much larger yield per hectare than wheat, the EU currently produces 2 million more tons of sugar beet than wheat on 20 million less hectares of land. Additionally, sugar beets produce more ethanol per hectare: a hectare of sugar beets can produce 30 hectoliters more ethanol, on average, than wheat. Also, sugar beet ethanol is shown to have a more energy-efficient production process than wheat ethanol. Currently, the most important bioethanol producers are France, Spain, Germany, Sweden, Poland, and Italy. In comparison with the USA and Brazil, EU ethanol for fuel production is still very modest. These 2 countries are now in competition for the title of world biggest producer. For both these countries, the main driver is reducing dependency from foreign fossil fuel sources.

Some excerpts taken from Bajpai (2007). PIRA Technology Report on Bioethanol with kind permission from Smithers PIRA

P. Bajpai, *Advances in Bioethanol*, SpringerBriefs in Applied Sciences and Technology, 79
DOI: 10.1007/978-81-322-1584-4_7, © The Author(s) 2013

Table 7.1 World fuel ethanol production

Continent	Millions of gallons
North & Central America	13,768
South America	5,800
Brazil	5,577
Europe	1,139
Asia	952
China	555
Canada	449
Australia	71
Africa	42

Based on RFA (2012)

In Europe, except for Sweden, and unlike the United States or Brazil, ethanol is not incorporated directly but transformed into ETBE (obtained by reacting isobutene, a liquefied petroleum gas, with ethanol) before being blended with gasoline. One reason for this regional particularity is the obligation to properly account for motor fuel properties such as volatility, since pure ethanol makes ethanol/gasoline blends more volatile. Another advantage of this practice is that it avoids separation of the alcohol and "gasoline" phases in the presence of traces of water.

7.2 Australia

The Australian government has supported ethanol since 2000 with a range of tax exemptions and production subsidies, aiming to produce 92 million gallons of biofuel by 2010, enough to replace 1 % of total fuel supply. 2006 production of ethanol stood at 39.4 million gallons. In view of low consumer confidence in ethanol, the government launched a campaign to encourage greater use of fuel ethanol. There is additional support at regional level in certain States (including Queensland), e.g., for biofuel marketing, the development of blending and distribution facilities and technical support for biofuel use. In Australia, bioethanol is produced from grain and sugarcane molasses. Australia only produces small amounts of biodiesel to date. However, several industrial scale plants are on the drawing board or in the construction phase. The fuel ethanol industry is still waiting for the big breakthrough, but political support for the industry is growing. Australia is a net importer of crude oil, and biofuels are considered to be increasingly important in view of surging world oil prices. Advocates of ethanol production also cite benefits for the ailing domestic sugarcane industry.

7.3 China

China is already the world's third largest producer of ethanol (90 % from corn) and has ambitious future growth targets for bioethanol from second generation waste biomass (Liu 2006; Zhenhong 2006). Current Chinese targets for bioethanol

Table 7.2 Ethanol Industry Overview

Year	Jan 99	Jan 00	Jan 01	Jan 02	Jan 03	Jan 04	Jan 05	Jan 06	Jan 07	Jan 08	Jan 09	Jan 10	Jan 11	Jan 12	Jan 13
Total Ethanol Plants	50	54	56	61	68	72	81	95	110	139	170	189	204	209	211
Ethanol Production Capacity (mgy)	1701.7	1748.7	1921.9	2347.3	2706.8	3100.8	3643.7	4336.4	5493.4	7888.4	10,569.4*	11,877.4	13,507.9	14,906.9	14,712.4
Plants Under Construction/ Expanding	5	6	5	13	11	15	16	31	76	61	24	15	10	2	4
Capacity Under Construction/ Expanding (mgy)	77	91.5	64.7	390.7	483	598	754	1778	5635.5	5536	2066	1432	522	140	158
States with Ethanol Plants	17	17	18	19	20	19	18	20	21	21	26	26	29	29	28

* operating plants
Based on RFA (2012)

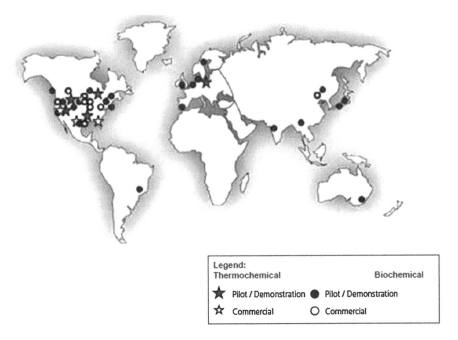

Fig. 7.1 Global and Lignocellulosic Ethanol facilities. Johnson et al. (2010), reproduced with permission

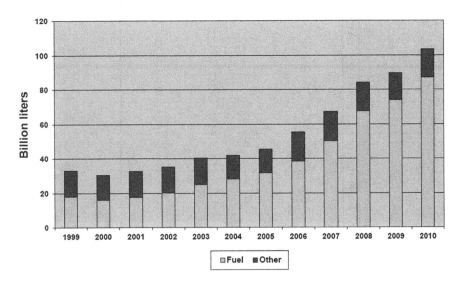

Fig. 7.2 Ethanol production by type. RFA 2010, reproduced with permission

(10 million tons by 2020) are considered conservative (Yan et al. 2010). Current bioethanol plants in China employ corn, wheat, and cassava, but sweet sorghum and sugarcane have future potential. Regarding second generation feedstocks, COFCO (China National Cereals, Oils and Foodstuffs Corporation) is investing 50 million Yuan (US$6.5 million) to build a cellulosic ethanol pilot plant in Zhaodong, in the northeastern province of Heilongjiang, with an annual capacity of 5,000 t.

Another cellulosic ethanol pilot plant with a production capacity of 10,000 t is being planned in the Yucheng area of Shandong (see http://www.biofuels.apec.org). So far, Chinese government has made mandatory the use of E10 in nine provinces (at central and northern China) that account for about one-sixth of that country's vehicles. Officials say that this mandate aims at reducing the oil demand—that is currently 40 % imported—and also aims at improving air quality in big cities. However, Chinese authorities also frequently highlight the targets of helping stabilize grain prices and raising farmer's income.

In China, more than 80 % of ethanol is produced from grains (corn, cassava, rice, etc.), about 10 % from sugarcane, 6 % from paper pulp waste residue, and the remainder is produced synthetically. So far, no significant amounts of biodiesel have been produced in China. There are no industrial scale biodiesel plants in the country. Fuel ethanol is exempted from consumption tax (5 %) and value-added tax (17 %).

7.4 United States

The USA has the world's fastest growing and largest fuel ethanol market. From its beginnings in the Midwest three decades ago, the US ethanol industry has grown to 211 plants operating in 29 states, with annual capacity of 14.8 billion gallons. The production of 13.3 billion gallons of ethanol in 2012 created real, measurable economic opportunity, including 70,000 direct jobs; 295,000 indirect and induced jobs; and $40.6 billion contribution to GDP' $28.9 billion in household income. Using the newest version of the GREET model developed by the US Department of Energy, the 13.3 billion gallons of ethanol produced in 2012 reduced greenhouse gas (GHG) emissions from on-road vehicles by 33.4 million tons. That's equivalent to removing 5.2 million cars and pickups from the road for one year. Table 7.3 shows Historic US fuel Ethanol Production data. Figure 7.3 shows Historic US ethanol import and exports.

Since 2006, fuel ethanol is consumed across the country and blended in 30 % of the gasoline consumed in USA (RFA 2007). Earlier, ethanol was used in niche markets in the Midwest, where the production is still concentrated. It is expected that fuel ethanol would be blended in 40 % of the gasoline consumed. Ethanol is sold in most states as octane enhancer or oxygenate blended with petrol and currently covers close to 3 % of the USA's gasoline demand. In recent years, fuel ethanol demand has been stimulated by the phasingout of MTBE as octane enhancer. A major concern with MTBE is water contamination and its health effects.

The cost of production of fuel ethanol from corn in USA in 2006 was estimated in the 0.33–0.50 Euro/l range against 0.21–0.29 Euro/l for cost of production from

Table 7.3 Historic US fuel
ethanol production

Years	Millions of gallons
1980	175
1981	215
1982	350
1983	375
1984	430
1985	610
1986	710
1987	830
1988	845
1989	870
1990	900
1991	950
1992	1,100
1993	1,200
1994	1,350
1995	1,400
1996	1,100
1997	1,300
1998	1,400
1999	1,470
2000	1,630
2001	1,770
2002	2,130
2003	2,800
2004	3,400
2005	3,904
2006	4,855
2007	6,500
2008	9,000
2009	10,600
2010	13,230
2011	13,900
2012	13,300

Based on RFA (2012)

sugarcane in Brazil (Worldwatch Institute 2006). The energy balance of ethanol production from corn is also much less favorable (1.34, according to Shapouri et al. 1995) than in Brazil (8.3–10, according to Macedo et al. (2004)). The US Department of Energy and the National Corn Growers Association are cooperating to promote the development of refueling infrastructure for E85 and to encourage fleet operators to choose ethanol to meet the alternatively fueled vehicles requirements of the Energy Policy Act. As previously stated, E85 blends require FFT. The potential phaseout of MTBE and an increasing emphasis on domestic energy supply and energy security are likely to favor increased use of fuel ethanol in the US.

Despite the strong interests of corn producers, the long-term sustainability of fuel ethanol production in USA will ultimately depend on the use of new

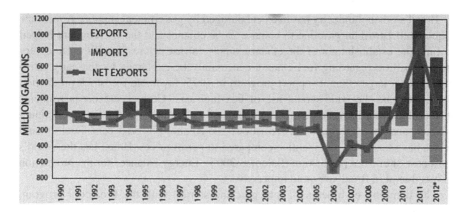

Fig. 7.3 Historic US ethanol import and exports. RFA (2013), reproduced with permission

feedstocks and, in this sense, there is a strong commitment to develop new routes of liquid fuels production from cellulosic material.

7.5 Brazil

Brazil has been for decades the world's largest producer and consumer of fuel ethanol, but was overtaken by USA in 2006. Brazil is the world's top exporter; in 2006, it exported 3.43 billion liters of ethanol. Brazil was the first country to embrace large-scale bioethanol production, via their government's *Proalcool* program that was initiated in 1975 to exploit sugarcane fuel alcohol as a gasoline substitute in response to rising oil prices. Brazil is now the world's second biggest producer with around 30 billion liters/annum (2008) from sugarcane and is the world's biggest exporter of fuel ethanol. The number of sugarcane bioethanol plants in Brazil will increase to over 400 in the next few years, and production is expected to reach 37 billion liters/year (from 728million tons of sugarcane) by 2012–2013 (Amorim et al. 2009; Basso and Rosa 2010). In Brazil, ethanol blends are mandatory (E20–E25) and anhydrous ethanol is also available from thousands of filling stations. In addition, there are 6 million flex-fuel vehicles in Brazil and 3 million able to run on E100. Bioethanol now accounts for ~50 % of the Brazilian transport fuel market, where gasoline may now be regarded as the "alternative" fuel.

In March 2007, Brazil and the United States signed a broad agreement to work together to advance biofuels technology and set common standards for ethanol trade. Brazil plans to double its ethanol production to 9.2 billion gallons per year and triple its ethanol exports to 2.6 billion gallons per year over the next seven years and will require about $13.4 billion in investments to meet that goal, Brazilian Agriculture Minister Luis Carlos Guedes Pinto said in an interview with Bloomberg published March 14, 2007.

Brazil's ethanol industry has been encouraged by government support. A mandate requires all Brazilian gasoline to contain 25 % ethanol and the Government helped to fund the establishment of the ethanol supply infrastructure. Brazil's current very strong position regarding fuel ethanol can mainly be explained by the early introduction of a large ethanol support program in 1975, favorable feedstock production conditions and the wide-spread use of flex-fuel cars. Current support instruments include blending provisions, minor mineral oil tax reductions for fuels containing ethanol, and motor vehicle tax reductions for ethanol-powered cars. Fuel ethanol prices have been very competitive compared to petrol prices. Brazil has the world's least cost of ethanol production (Salomao 2005). Energy balances for the production of bioethanol in Brazil are quite impressive.

The blending of bioethanol into gasoline started with the Proálcool program after the first oil crisis in 1975. Today, there is no ethanol-free gasoline on the Brazilian fuel market. All gasoline is marketed with a 25 % share of bioethanol (E25, also called gasohol. Actual ethanol content varies as it is yearly adapted to the market situation) and pure bioethanol is on the market as well.

Brazil is the only country in the world with the conditions to quite considerably expand its capacity of ethanol production in short- to mid-term.

7.6 Canada

Canada currently produces relatively low volumes of fuel ethanol. A number of initiatives are underway to boost production significantly. In 2006, total fuel ethanol production amounted to about 150 million gallons. Production is expected to increase significantly in the next few years if current and announced biofuels programs are implemented. To meet Kyoto Protocol commitments, the country aims to replace 35 % of its gasoline use with E10 blends, requiring production of 350 million gallons of ethanol. Seven new plants with total capacity of 200 million gallons are planned under the Ethanol Expansion Program. Ontario, Saskatchewan, and Manitoba are already promoting ethanol through production subsidies, tax breaks, and blending requirements. In Canada, ethanol is produced almost entirely from cereals. The company Iogen Corp. maintains a demonstration plant for producing ethanol from cellulose in Ottawa, but this plant produces demonstration, not commercial, quantities of cellulose ethanol (Iogen 2005).

Iogen's first 90 ML/yr commercial plant is planned for Birch Hill, Saskatchewan, an area with sustainable supplies of straw and green residues. The availability of straw, combined with government support, were key factors in this decision. This will be first facility to take advantage of the NextGen Biofuels fund. Iogen has "suspended" its 68 ML/yr plant in Shelley, Idaho.

Ethanol-blended gasoline is now available at over 700 gas bars across Canada from Quebec to the Pacific, including the Yukon Territory. In many regions, ethanol blends are available for bulk delivery for farm and fleet use. The federal government and several provinces offer tax incentives.

7.7 India

India is the second largest producer of ethanol in Asia and is one of the world's largest sugar producers. In February 2009, India and the US exchanged a memorandum for cooperation on biofuels development, covering the production, utilization, distribution, and marketing of biofuels in India. Traditionally most of the Indian ethanol production is directed to the industrial consumption. Recently, mainly due to economic and strategic reasons, Indian government has seriously considered fuel ethanol production and a mandate for E10 blends is currently effective in 13 states. In the near future, the mandate for E10 shall be effective in the whole country. In addition, the Indian Institute of Petroleum had conducted experiments using a 10 % ethanol blend in gasoline and 15 % ethanol in biodiesel (DSD 2005).

7.8 Thailand

The Thai government is pursuing a policy of increasing consumption of biodiesel and ethanol produced domestically from cassava. One of its initial steps is to replace the octane-enhancing additive MTBE in gasoline with ethanol. Large-scale production of fuel ethanol production has started with molasses, but cassava was officially designated the prime raw material. Thailand is a large producer of cassava. As the domestic price of sugar is too high it doesn't seem worth to produce ethanol from sugarcane. In Thailand, all premium gasoline shall be replaced by E10.

7.9 Japan

Japan is one the main consumers of motor gasoline in the world and is heavily dependent on imported oil (Ethanol News 2006). The country has considered large-scale use of fuel ethanol, or ETBE, targeting at improving its energy security and at reducing GHG emissions, in this case in order to accomplish its Kyoto obligations. By now ethanol blends are used in some regions. Japanese government intends to define a mandate for E3 valid for the whole country and to expand this mandate to E10 in next few years. However, there is some resistance due to the low number of large-scale ethanol suppliers and also due to the interests of oil companies that prefer gasoline blends with ETBE rather than with fuel ethanol (Piacente 2006).

References

Amorim, HV, Basso, LC, Lopes, ML (2009) Sugar can juice and molasses, beet molasses and sweet sorghum: composition and usage. In: The alcohol textbook, 5th edn. Nottingham University Press, Nottingham, pp 39–46

Bajpai P (2007) Bioethanol. PIRA Technology Report, Smithers PIRA, UK

Basso LC, Rosa CA (2010) Sugar cane for potable and fuel ethanol. In: Proceedings of the worldwide distilled spirits conference, Edinburgh. Nottingham University Press, Nottingham

DSD (2005) Dutch Sustainable Development Group. Feasibility study on an effective and sustainable bio-ethanol production program by Least Developed Countries as alternative to cane sugar export

Ethanol News (2006). Japan to fight global warming, rising oil prices by replacing gas cars with ethanol ones. Available at http://ethanol-news.newslib.com/story/6938-4598/69

Iogen Corporation (2005) Iogen corporation. http://www.iogen.ca/company/about/index.html

Johnson T, Johnson B, Scott-Kerr C, Kiviaho J (2010). Bioethanol—status report on bioethanol production from wood and other lignocellulosic feedstocks. In: 63rd appita annual conference and exhibition, Melbourne, 19–22 April 2009

Liu D (2006) Chinese development status of bioethanol and biodiesel. In: World biofuels symposium, Beijing, China

Macedo, IC, Leal, MRLV, Silva, JEAR (2004) Assessment of greenhouse gas emissions in the production and use of fuel ethanol in Brazil. Secretariat of the Environment, State of São Paulo

Piacente EA (2006) Perspectives for Brazil in the bio-ethanol market. MSc dissertation. State University of Campinas—Unicamp, Campinas

Renewable Fuels Association (RFA) (2007) Ethanol industry outlook: building new horizons. Available at: http://www.ethanolrfa.org/objects/pdf/outlook/RFA_Outlook_2007.pdf

Renewable Fuels Association (RFA) (2010) http://www.ethanolrfa.org/pages/statistics

Renewable Fuels Association (RFA) (2012) http://www.ethanolrfa.org/pages/statistics

Renewable Fuels Association (RFA) (2013) Battling for the barrell. ethanolrfa.3cdn.net/dc207800043a5aa5aa_y5im6rokb.pdf

Salomao A (2005) O novo siclo da cana de acucar. Exame, ed. 845 No. 12, 22 June

Shapouri H, Duffield JA, Graboski M S (1995) Estimating the net energy balance of corn ethanol. AER-721. USDA Economic Research Service, Washington, D.C

Yan X, Inderwildi OR, King DA (2010) Biofuels and synthetic fuels in the US and China: a review of well-to-well energy use and greenhouse gas emissions with the impact of land-use change. Energy Environ Sci 3:190–197

Zhenhong Y (2006) Bio-fuels industry in China: utilization of ethanol and biodiesel in today and future. In: World biofuels symposium, Beijing, China

Chapter 8
Future of Bioethanol

The future of bioethanol appears to be bright as the need for renewable energy sources to replace dependence on foreign oil is in high demand. With many nations seeking to reduce petroleum imports, boost rural economies, and improve air quality, world ethanol production rose to 13,300 million gallons in 2012. The success of domestic ethanol industries in the USA and Brazil has sparked tremendous interest in countries across the globe where nations have created ethanol programs seeking to reduce their dependence on imported energy, provide economic boosts to their rural economies, and improve the environment. As concerns over greenhouse gas emissions grow and supplies of world oil are depleted, Europe and countries like China, India, Australia, and some Southeast Asian nations are rapidly expanding their biofuels production and use.

A lot of research is being done including turning biomass, materials from plants, into ethanol using special biotechnological methods. Biomass ethanol is the future of ethanol production because biomass feedstocks, like wheat straw or switchgrass, require less fossil fuels to grow, harvest, and produce. It also allows us to utilize more marginal land, such as grasslands, rather than precious acreage devoted to food crops like corn or soybeans. In this way, ethanol production from biomass does not negatively affect the livestock and food industry. The biorefinery, analogous to today's oil refineries, could economically convert lignocellulose to array of fuels and chemicals—not just ethanol by integrating bio- and thermochemical conversion. Fundamental research and partnerships with the emerging bioenergy industry are critical for the success.

There has been continuing research on improving the energy output of ethanol and improvements should keep growing. Right now, more and more E85 stations are popping up everywhere and more products from generators to power tools to lawn mowers will all start to use some alternative fuels. There are already engines that can run 100 % pure ethanol and improvements will help migrate these engines

---

Some excerpts taken from Bajpai (2007). PIRA Technology Report on Bioethanol with kind permission from Smithers PIRA

P. Bajpai, *Advances in Bioethanol*, SpringerBriefs in Applied Sciences and Technology, DOI: 10.1007/978-81-322-1584-4_8, © The Author(s) 2013

to other areas. Big auto manufacturers like Nissan, Ford, and Honda have all invested money into E85 models as well. Portable generators, standby and emergency generators should all start using ethanol as a fuel source as well. Hopefully an alternative fuel like ethanol will be more popular before the big boom in India and China car usage begins to lower environmental pollution.

The emergence of carbon-trading programs in response to many countries' ratification of the Kyoto Protocol will also enhance the affordability of ethanol fuels in comparison to gasoline and diesel. Because ethanol fuels offer a substantial reduction in carbon dioxide emissions, users can obtain carbon credits that can be sold to heavy polluters, again reducing ethanol costs while increasing that of fossil fuels. The European Union recently developed a carbon-trading program while Japan has conducted several scenario simulations, with hopes to initiate its own nationwide trading system. As Russia considers ratification of the Kyoto Protocol, which would bring the agreement into effect, it seems likely that similar carbon-trading schemes will continue to emerge around the world.

A combination of well-reasoned government policies and technological advancements in ethanol fuels could guide a smooth transition away from fossil fuels in the transportation sector. As environmental externalities continue to be incorporated into policy consideration and the fledgling industry emerges, ethanol fuels are likely to become an increasing attractive fuel alternative in the foreseeable future. Looking into the future, the ethanol industry envisions a time when ethanol may be used as a fuel to produce hydrogen for fuel cell vehicle applications.

Reference

Bajpai P (2007) Bioethanol. PIRA Technology Report, Smithers PIRA, UK

About the Author

Dr. Pratima Bajpai is currently working as a Consultant in the field of Paper and Pulp. She has vast experience of 36 years in this field. She has worked at National Sugar Institute Kanpur, University of Saskatchewan and University of Western Ontario in Canada and Thapar Center for Industrial Research and Development in India. Dr. Bajpai's main areas of expertise are industrial biotechnology, pulp and paper and environmental biotechnology. She has immensely contributed to the field of industrial biotechnology and is a recognized expert in the field. Dr. Bajpai has written 12 advanced level technical books on environmental and biotechnological aspects of pulp and paper, which have been published by leading publishers in the USA and Europe. She has published 7 book chapters, 3 chapters in encyclopedias, 87 articles in international peer reviewed journals and 52 papers in conferences and seminars. She holds 7 patents in her name and 4 applications are pending. Dr. Bajpai has handled 33 sponsored research projects from industry and government agencies. She is an active member of New York Academy of Science, American Society for Microbiology, and many more.

P. Bajpai, *Advances in Bioethanol*, SpringerBriefs in Applied Sciences and Technology,   91
DOI: 10.1007/978-81-322-1584-4, © The Author(s) 2013

Index

A

Acid hydrolysis, 33, 34
Adsorption, 34, 35
Agricultural waste, 15, 24
Air toxins, 69
Alcohol, 5, 7, 13, 15, 19, 28, 29, 44, 65, 67, 71, 72, 77, 80, 84
Alcohol dehydrogenase, 36
Alfalfa, 16, 17
Alginate, 17
Alginic acid, 17
Alpha amylase, 28
Alternative energy, 1
Alternative fuel, 1, 7, 89, 90
Ammonia fiber explosion, 32
Amylase, 30, 34
Amyloglucosidase, 30
Amylolysis, 26
Amylopectin, 29
Anhydrous ethanol, 2, 15, 57–59, 84
Animal feed, 28, 30, 41
Arabinose, 36, 37
Automobile, 7, 13, 55, 58, 59, 62, 65, 68, 71, 75
Aviation, 2, 57, 62, 64

B

Bacteria, 18, 28, 34, 36, 37, 38, 45
Bagasse, 15, 17, 44, 47, 48
Barley, 15, 30, 44
Betaglucosidase, 34
Beverage, 4, 5, 7, 14, 15, 29, 30
Beverage waste, 19
Biodiesel, 6, 7, 9, 10, 18, 64, 81, 82, 86

Bioethanol, 2, 3, 9, 10, 15, 17, 18, 25, 26, 28, 30, 32, 36, 41, 42, 44, 46, 47, 56–58, 79, 81, 84, 85, 89
Biofuel, 1, 6, 8–10, 11, 15, 17, 42, 43, 46, 47, 55, 79–81, 84–86, 89
Biological treatment, 32
Biomas, 15, 17, 18, 81, 89
Bioprocessing, 38, 47
Boat, 65
Brewery waste, 19

C

Carbohydrates, 17, 27, 33
Carbon dioxide, 1, 2, 14, 28, 38, 58, 69, 76, 77, 90
Carbon dioxide emissions, 69, 70
Carbon monoxide, 56, 65, 68–70
Carbon monoxide emission, 68, 69
Carboxylic acid, 33
Cassava, 1, 81, 86
Cellobiohydrolase, 34
Cellobiose, 34, 37
Cellulase, 34, 35, 37, 38, 40
Cellulolytic enzyme, 37
Cellulose, 5, 17–19, 21, 24, 32–35, 37, 38, 42, 85
Cereal, 24, 26, 28, 85
Chainsaw, 65
Chitin, 18
Combustion engine, 2, 14, 62, 67, 75
Condensed distillers solubles, 29
Corn, 1, 3, 6, 9, 10, 15, 17, 19, 21, 26–28, 30, 40–42, 44–48, 55, 62, 69, 81–84, 89
Corn cob, 1